Gliding To A Better Place

Profiles From Ohio's Territorial Era

Gary S. Williams

Buckeye Books
Caldwell, Ohio
2000

Copyright © 2000 by Gary S. Williams
All rights reserved.

Permission to reproduce in any form
must be secured from the author.

Illustrations by Owen Williams

Please direct all correspondence and book orders to:
Gary S. Williams
42100 Township Road 491
Caldwell, OH 43724
Phone: (740) 732-8169
e-mail: buckeyebooks@hotmail.com

Library of Congress Control Number 00-92175
ISBN 0-9703395-0-X

Published for Buckeye Books by
Gateway Press, Inc., 1001 N. Calvert Street
Baltimore, MD 21202

Printed in the United States of America

TABLE OF CONTENTS

INTRODUCTION		v
CHAPTER I:	George Washington Slept Here	1
CHAPTER II:	John Gibson: Unsung Hero	18
CHAPTER III:	John Heckewelder: A Bridge Between Worlds	49
CHAPTER IV:	Anne Bailey: Lady of Legend	73
CHAPTER V:	James Wilkinson: They Weren't All Heroes	88
CHAPTER VI:	Pontiac, Joseph Brant, Tecumseh: A Trio of Natives	118
CHAPTER VII:	Abraham Whipple: Last Voyage of the Commodore	160
CHAPTER VIII:	William Maxwell and Nathaniel Willis: Pioneer Printers	181
CHAPTER IX:	Benjamin Tappan: The Curmudgeonly Politician	195
CHAPTER X:	John Chapman: The True Story of Johnny Appleseed	214
BIBLIOGRAPHY		237
INDEX		243

For Mary

INTRODUCTION

Ohio is the start of it all when it comes to U.S. westward expansion. Not only was Ohio the first place settled by citizens of the United States, it may well be the first place in the world founded by people with government mandated civil liberties.

All of the states that entered the Union before Ohio--the original 13 plus Vermont, Kentucky and Tennessee-- had been originally settled by British subjects. As the first new area to be settled after independence, Ohio was the first step on the road to Manifest Destiny. What happened here set a precedent for all future expansion.

At about the same time the Constitutional Convention was meeting in Philadelphia to provide a permanent framework for the new government, Continental Congress was holding its last session in New York. Before this unwieldy system of government was retired, it established the Northwest Territory and provided for its governance with the Ordinance of 1787. This document also included a modified Bill of Rights which prohibited slavery and established freedom of religion and certain legal rights. Most states had some sort of Bill of Rights, and the federal government did after 1791, but when the original pioneers landed at Marietta in 1788, they were the first people to settle new lands with guaranteed civil liberties emanating from federal authority.

The issue of western lands was one of the most important to be addressed by the untested republic, and expansion into territory held by Native Americans that bordered lands claimed by European powers was a perilous quest that held great implications for the nation's future. Opportunists of every stripe came flooding westward to meet this challenge. The motto on the seal of the newly established Territory Northwest of the River Ohio was *Meliorem Lapsa Locavit,* which can be roughly translated from the Latin as "Gliding to a Better Place." The hopeful pioneers who arrived here via the Ohio River or Lake Erie were willing to work to make this a better place.

What sort of person was it that came gliding to Ohio 200 years ago? To begin with, they were a diverse lot from all over. The people profiled in this book were born in six different colonies and England, which makes a case for Ohio being the nation's first melting pot. The characters discussed here are some of the more interesting people who played a role in Ohio's Territorial Period. A mix of schemers and saints, soldiers and settlers, among them they were involved in every important phase of Ohio's history before statehood, and their collective story could serve as a general history of the era.

Some of the figures included here are nationally known, but not neccesarily for their Ohio connection. Not only did George Washington visit

Ohio, but as a militia officer at age 20 he started a world war that wrested the entire Ohio Valley from the French. In doing this, he became the first man to bring wheeled vehicles and artillery across the Appalachian Mountains, and the route he made became the National Road. All schoolchildren are familiar with the legend of Johnny Appleseed, but not everyone knows this legend was a real person who spent most of his adult life in Ohio preparing the way for the pioneers. And of the Native Americans who tried to unify the diverse tribes in the face of white encroachment, it is not known that the three most significant - Pontiac, Joseph Brant, and Tecumseh - were all born in what is now Ohio.

Other chapters are devoted to lesser known but significant and often colorful characters who played a role in the state's development. Among this second tier of early Ohio heroes are such people as: pioneer printers William Maxwell and Nathaniel Willis, whose primitive presses paved the path to statehood; "Mad Anne" Bailey, a hard-living frontier scout as tough as any man; Western Reserve pioneer and curmudgeonly politician Benjamin Tappan; John Gibson, who played a prominent role in nearly every frontier conflict in a 60 year period; and Abraham Whipple, a naval hero turned farmer who at age 68 became the first man to sail a ship down the Ohio and Mississippi Rivers.

In terms of character, these people range from the saintly Moravian missionary John Heckewelder to the villainous traitor James Wilkinson. The wide open frontier was often a small world, and many of those profiled here knew each other as well as they did the other well known figures of early Ohio history. Among them all, they forged a state out of the wilderness and paved the way for all who followed. The story of these people who came gliding to a better place is the story of America itself.

ACKNOWLEDGMENTS

It's been a lot of fun doing the research for this book, and in my travels that ranged from Johnson Hall to Grouseland, I met a lot of helpful people. Librarians and museum staff were supportive everywhere, but I had particularly encouraging encounters at the Campus Martius Museum in Marietta, the Fort Recovery Museum, Tu-Endi-We Park in Point Pleasant, West Virginia, and Grouseland Museum in Vincennes, Indiana.

I would also like to thank the following: Bob Hirche, for enthusiastic encouragement and advice; Alan Marcosson, for serving as a sounding board; Myrtle Ake, who got me started on my first chapter; Barbara Lyons of Ohio State University Press, who offered kind words at a key time; Earl Olmstead, who shared research tips from his own books; and Roger Pickenpaugh, who gave valuable advice at every stage based on his experiences with his books.

But this book has mainly been a Williams family effort. Thanks go to my parents, from whom I gained an appreciation of both writing and history; my brother Drew, who did the map; my brother Brian, who supervised the editing process; my daughter Meryl, who helped with the typing; and my son Owen, who did the illustrations. Most of all, it was my wife Mary, who patiently did all layout and formatting, offered essential writing and editing tips and objective assessment, and supported me completely in the pursuit of my dreams. The phrase "I couldn't have done it without you" was just a cliche before.

OHIO
Before Statehood

CHAPTER I
George Washington Slept Here

On a cold, cloudy evening in late October, two large canoes came floating down the Ohio River. The ten occupants pulled up on the Ohio side and set up camp for the night. These men noticed an abundance of deer, wild turkey and other game, but this was not a hunting expedition. Still, they were not averse to a little night fishing. One member of the party was keeping a journal, and he wrote that "at this place we through (sic) out some Lines at Night and found a Cat fish the size of our largest River Cats hooked to it in the Morning, tho it was of the smallest kind here."

That diarist was George Washington, and his entry for October 25, 1770 was made near present-day Fly, Ohio. The 38-year-old Washington was on the most westward trip of his life, heading as far as the Kanawha River to scout out land to be made available as bounties for French and Indian War veterans. This expedition means that Ohioans can truly claim that George Washington slept here, as at least seven of his campsites are known to have been on the Ohio side of the River.

But more importantly, it serves as an illustration of Washington's involvement with the West and his role in the development of Ohio. Washington always looked to the West for the future, beginning with his service as a 20-year-old militia officer trying to remove the French from the Ohio River Valley. At that time, there were almost no English-speaking whites in the region, but when Washington left the Presidency over 40 years later, Ohio was on the verge of statehood. His involvement in making this change possible shows that the Father of our Country's sobriquet expanded beyond the thirteen colonies he liberated.

Although he was born on the Atlantic seaboard in 1732, Washington always had an eye on the West. He was a member of a landholding family, and to Virginians, land was a precious commodity, with wealth being measured in acres. As a member of the landed gentry, Washington could have been a wealthy idler, but he was instilled with a sense of *noblesse oblige*. George's father died when the boy was young, but his older half brothers Lawrence and Augustine saw that he got an education.

Young George was a poor speller and writer, but he had a talent for math and science. Through a family connection with Lord Fairfax, an English nobleman with extensive Virginia land holdings, Washington got involved in surveying. This seemed an ideal vocation as it satisfied the rugged young man's desire to be outdoors, utilized his math skills, and served the Virginian's interest in a profitable way. At the age of 16, he began accompanying surveying parties to western Virginia.

It was a dispute over the borders of western Virginia that got Washington involved in a new profession. As the possibility of conflict with France arose, Washington was appointed a major in the Virginia militia. Virginia claimed borders that ran far to the west and northwest, which brought them into conflict with not only Pennsylvania but with the French who were already trading in the area. The French explorer LaSalle had traveled the Great Lakes and the Ohio and Mississippi Rivers in the 1670's and had claimed the entire region for France.

The French, however, were interested in fur trade with the Indians rather than colonization, and they maintained only widely scattered posts throughout the Midwest. To reinforce their claim, in 1749, the French sent a force under Celeron de Blaineville that left lead plates at several major river junctions proclaiming jurisdiction. But as historian Dale Van Every noted, it would take lead "in a more potent form" to combat British traders who were encroaching upon French trade with cheaper goods. At Pickawillany, near present day Piqua, a Miami chief became known as Old Britain because of his preference for British goods and loyalty to their traders. In 1752, a party of Indians under French leadership attacked the village, scattered the traders, and boiled and ate Old Britain. The next year the French began constructing a series of forts that would give them a portage from Lake Erie to the beginning of the Ohio River.

The Virginians were quite alarmed by these developments, as they had plans to settle lands that drained into the Ohio. In 1748, the Ohio Company of Virginia had been chartered, with Lawrence and Augustine Washington as members. This group was offered large land grants provided they could produce settlers and build forts on the new lands. They hired experienced frontiersman Christopher Gist to explore their new dominions. Gist's overland trip through Ohio in 1750 was the first recorded visit to Ohio by an English-speaking white man.

France and England were clearly on a collision course in the Ohio Valley as the Virginia colonials moved up the Monongahela while the French were heading down the Allegheny. At the suggestion of Lord Fairfax, Virginia's royal governor Robert Dinwiddie made George Washington a major in the Virginia militia. Then he gave the 20-year-old officer the dangerous diplomatic mission of carrying a letter to the French forts asking them to vacate British lands. In carrying out this mission, Washington became in Van Every's words, "the first officially commissioned representative of his country to cross the mountains."

Washington set out from Logstown on the Ohio River, but with his military eye he saw that 20 miles upstream at the confluence of the Monongahela and Allegheny Rivers was a better location for a fort. Traveling with him were the 46-year-old Gist and a few Indians friendly to the British. In December they went up the Allegheny to Fort Venango, the current southernmost French post. The commander there felt unauthorized to respond and sent the messengers further north to Fort LeBoeuf. Here they were received and invited to wait while a response was prepared.

Washington used this opportunity to scout the condition of the fort and troops, while the French spent their time trying to bribe the Indians into joining them. After a few days, Washington was given a polite letter of refusal to take back. The return trip was a perilous one, as Washington was shot at by Indians, caught in snowstorms, and thrown from a raft into the ice-choked river. But he safely returned to Williamsburg in January, and Dinwiddie tried to generate publicity for the impending conflict by publishing the young major's journal of his trip.

At Washington's urging, a force was sent to build a fort at the forks of the Ohio before the French could. Washington was promoted to Lieutenant Colonel and sent back in the spring of 1754 to reinforce this group. He left with 300 men and the first wheeled vehicles and artillery to cross the mountains. The route he made later became the National Road, the first highwy to the West.

Unfortunately, a large French force had already appeared at the forks of the Ohio and driven the Virginians off without firing a shot. Washington met this group on their retreat, and soon afterwards got word from his Indian scouts that a contingent of French troops was nearby.

The French later claimed that this was a diplomatic party like Washington's previous mission had been. But as it was a large armed party traveling furtively by avoiding the main paths, Washington decided their intent was hostile and resolved to attack. On the morning of May 28, the Virginians and their Indian allies surprised the French and started a world war. The incident also launched a new career for Washington, who found something he liked better than surveying. He wrote of his first battle, "I have heard the whine of bullets, and believe me, there is something charming in the sound."

Washington's first battle was a complete success, as the French had several killed, including their leader, and the rest taken prisoner. But the victory was short-lived and had disastrous repercussions. The main French army was not far away, and his tired and poorly provisioned troops were a long way from their base of supply. When promised reinforcements did not arrive, Washington was forced to retreat towards Cumberland, Mayland.

But his troops could go no further than the Great Meadows near the Youghiogheny River. Here necessity forced them to build a crude circular fort that was called Fort Necessity. The Indians were perceptive of the situation, and when they realized the plight of the Virginians, they deserted almost to a man and melted back into the forest. The French surrounded the fort and began firing just as an all-day rain began. Washington tried to put up a fight, but the situation was hopeless, and after a parlay, he agreed to surrender the muddy fort just after midnight on July 4, 1754.

The defeated garrison was permitted to return home, but Washington and his men were vilified in France. In signing the terms of capitulation, Washington, who didn't read French, admitted to the "assassination" of the French leader on May 28. The two nations had been on a collision course, but as Voltaire said, "a cannon shot fired in the woods of America was the signal that set Europe ablaze." Within two years nearly all of Europe was at war, and a result of the war that Washington started was that England was finally able to remove the French from North America.

But there were several setbacks along the way. In 1755, the British sent regular troops to the frontier to try and accomplish what the Virginia militia could not. General Edwin Braddock and two regiments attempted to capture Fort Duquesne which the French had erected at the forks of the Ohio. Braddock had little respect for the Americans--in fact Washington and Ben Franklin were about the only Americans he had any kind words for. But he did respect Washington's experience in the area and offered him a position as aide-de-camp. Braddock followed Washington's route and built a road towards Fort Duquesne. But Braddock's insistence on following European-style tactics lead to disaster when his army was ambushed within a few miles of the fort on July 8, 1755. The red-coated British were easy targets in the woods and a defeat turned into a route. With Braddock mortally wounded, Washington took charge during the retreat , and once again returned home with a defeated army.

It was three years before the British mounted another assault on Fort Duquesne. This time it was a force under General John Forbes that built another route by way of Pennsylvania. Washington argued against this plan, in part because he feared a Pennsylvania route would weaken Virginia's claim to the area. But Forbes succeeded as his patient strategy paid off when the French abandoned the post on November 25, 1758. The British rebuilt the fort and renamed it Fort Pitt. The next year Quebec fell to the British, and in the Treaty of Paris in 1763, France had to give up all her North American claims

Gliding to a Better Place

to England.

Six weeks after serving with Forbes, Washington married Martha Custis and settled down to become squire of Mount Vernon. For the next 12 years he remained there as a prominent planter and community leader. But he remained interested in western land speculation and kept in touch with many of his old comrades. One of these men was William Crawford, a friend and fellow surveyor who was about the same age as Washington. The two entered into a partnership as it became known that western lands would be made available for veterans of the French and Indian War.

In the fall of 1770 Washington journeyed to Crawford's western Pennsylvania home to begin an exploration of the available bounty lands. Accompanying them were Dr. Robert Craik (a friend of Washington and a veteran of Fort Necessity), several of Crawford's employees, and two Indian guides. The group left Pittsburgh on October 20 and floated down the Ohio in two large canoes.

They entered what is now Ohio the next day and were immediately struck by the abundance of game, as they killed five wild turkeys the first day. They camped near present-day East Liverpool that night, where it started snowing around midnight and kept it up most of the night. The next day they continued on to a large Mingo village located near present-day Mingo Junction. Here they were disturbed by rumors that two traders had been killed by Indians downstream. But this threat to their expedition turned out to be a wild exaggeration, as what had actually happened was a trader had drowned trying to ford the Ohio.

So they continued downstream, camping the next three nights near the present-day towns of Powhaten Point, Fly, and Reno. On October 28, they encountered a large Indian hunting party led by the Mingo chief Guyasuta. By a strange coincidence, Guyasuta had been one of the Indians who had gone with Washington on his trip to Fort LeBouef 17 years earlier.

Since then, Guyasuta had switched sides and had served with the French at Braddock's Defeat. In fact, he readily and cheerfully acknowledged trying to shoot Washington at that battle. Washington's clothes were hit by bullets and he had horses shot out from under him that day, but he was not hit, and Guyasuta had concluded that Washington had "a charmed life." Guyasuta had continued to oppose the British, and during Pontiac's Conspiracy he led the Indians who captured and massacred the garrison at Fort Venango. He also been one of the commanders of the force that tried to prevent Henry Bouquet's relief of Fort Pitt during Pontiac's Conspiracy. Later he would again be an enemy as he attacked American settlements in Pennsylvania during the Revolution.

But for now this once and future foe was an ally, and he was quite excited about seeing his old traveling partner again. Guyasuta proposed the two parties camp together, which they did, probably in what is now Meigs County. The Indians staged elaborate rituals for his benefit, which Washington endured politely. But he was eager to get back to the business at hand. In his journal he wrote that "the tedious ceremony with which the Indians observe in their Councellings and speeches detained us till 9 o' clock." They encountered this party again on their return upstream, and this time Washington complained "by the Kindness and Idle ceremony of the Indians, I was detained at Kiashuta's Camp all the remaining part of the day", although he conceded that he did have "a good deal of conversation with him on the Subject of Land."

Washington's party followed the Ohio as far as the mouth of the Kanawha at present-day Point Pleasant, West Virginia, and then spent a few days exploring up that stream before turning back. They continued scouting for land on the return trip, with Washington and Crawford also taking some overland side trips. But it is a wonder that Washington loved the Ohio Valley so much, since every time he was there he encountered terrible weather. He had to deal with blizzards in 1753, surrendered in a rainstorm in 1754, and now he had to deal with fall flooding on the Ohio. Steady rains had swollen all the streams in the valley, making upstream travel difficult, and the party abandoned the river at the Mingo village and returned to Pittsburgh via land, arriving on November 21.

Washington never returned to Ohio, but the area remained important to him. Although he started out a provincial in every sense of the word, by the time Continental Congress named him Commander of the American Army, Washington had become an American. His views had expanded from how the Ohio Valley could benefit his pocketbook and his colony to how the area could help the new nation. In the dark days of the Revolution he talked of escaping to the West and continuing the war from there. And when victory was in sight he realized that the western lands offered payment to his loyal troops and a future for the new nation.

Late in the war, several of his officers had gathered in Newburgh, New York, and passed a resolution asking for western land as payment. The Newburgh Petition was presented to Washington by General Rufus Putnam of Massachusetts, and the two friends continued to correspond about the issue after the war.

A problem with allocating western lands was not only that Indians were currently occupying them, but also that several individual colonies had conflicting claims on them. These conflicts had to be resolved before settlement could occur, and the process by which they were resolved mirrored the Americans' coming together as a nation.

During the Revolution, the colonies were governed by the Articles of Confederation, which was a loose and weak agreement where any one state could veto any action. This unwieldy system needed

Gliding to a Better Place

to be replaced by a more permanent and stronger form of national government. But before this system was retired, it made a couple of significant accomplishments.

In 1785, Continental Congress passed a Land Ordinance that provided for survey and sale of western lands, but not their governance. With the foundations for legal settlement established, the Ohio Company of Associates was formed in Boston on March 1, 1786. Many of Washington's former officers were prominent in this settlement plan, and among the most impressive of these was Manasseh Cutler. Among the careers of the multi-talented Cutler were whaler, lawyer, minister, doctor, scientist, lobbyist and congressman. It was as a lobbyist extraordinaire that he played a major role in the development of Ohio.

In 1787 he was able to guide through Continental Congress passage of the Northwest Ordinance, one of the last and most significant acts of that body. This revolutionary document provided for governance of the new Northwest Territory, set the pattern that all future states would follow, and provided a seminal Bill of Rights.

The Ordinance of 1787 spelled out the process of attaining statehood, and stated that the initial territorial government would consist of a governor, a secretary, and three judges, or legislators. Another article encouraged education and resulted in an early emphasis on learning. It is worth noting that Ohio University, established in

1804 and graduating its first class in 1815, is older than Thomas Jefferson's University of Virginia.

The final articles of the Ordinance guaranteed freedom of religion and right to trial by jury, and forbade slavery. Many state legislatures had passed bills of rights during the Revolution, but this was the first one emanating from any national authority. This means that Ohio was one of the first places in the world organized and settled by free people with guaranteed liberties.

George Washington was not present for the passage of the Ordinance of 1787 because at the time he was presiding over the Constitutional Convention in Philadelphia. Here the greatest minds in the country were trying to provide the foundation for a permanent nation. Among the many crucial items to be addressed was the admittance of future western states to the union. After much discussion, they decided that future states would be admitted on an equal footing with the original thirteen.

So the road to Manifest Destiny began in Ohio. The original states plus the first additions of Vermont, Kentucky, and Tennessee had all been settled by British colonists. Ohio was the first place settled by citizens of the United States, and what happened here set a precedent for all future expansion. Now that there were the protections of the U. S. Constitution and progress had been made on relinquishment of state land claims, the U.S. westward movement could begin.

Gliding to a Better Place

The Ohio Company was able to purchase one and a half million acres near the Muskingum River in October of 1787, and many of the company leaders were also named to prominent positions in the new Northwest Territory. The New Englanders of the Ohio Company were an impressive lot, and many of their members had college backgrounds. Cutler was a Yale graduate, Territorial Secretary Winthrop Sargent and Territorial Judge Samuel Parsons were both Harvard men, and Territorial Judge James Varnum was an alumnus of Brown. There also were a lot of experienced veterans of Washington's army, with Parsons, Varnum, Ohio Company Superintendent Rufus Putnam and Territorial Governor Arthur St. Clair all having attained the rank of general.

Washington himself could not have given a stronger endorsement of the proposed settlement at Marietta when he wrote, "No colony in America was ever settled under such favorable auspices as that which has just commenced on the banks of the Muskingum. Information, property, and strength will be its characteristics. I know many of the settlers personally, and there never were men better calculated to promote the welfare of such a community."

The Ohio settlers wintered in western Pennsylvania in 1787 and built boats to float down the Ohio in the spring. On April 7, 1788, the original 48 pioneers landed at the mouth of the Muskingum, and immediately showed their respect for law and order by posting their first rules on a tree. Governor St. Clair arrived at the Marieta settlement on July 9 and inaugurated the government of the Northwest Territory. His first order of business was to negotiate with area tribes for land concessions.

Here the new governor had to work with the Washington administration in delicate diplomatic matters. As expected, Washington had been elected President after the Constitution had been ratified, and he took office in April of 1789. The issue of western land was one of the most important to be addressed by the untested republic, and expansion into territory held by Native Americans that bordered lands claimed by European powers was a perilous quest that held great implications for the nation's future.

Once the Indians realized the extent of the American threat, their resistance hardened. General Josiah Harmar , who commanded the tiny U. S. Army, was sent to subdue the Indians but succeeded only in stirring them up more. Governor St. Clair resolved to lead the next expedition himself, and Washington warned his General to beware of a surprise attack. When St. Clair's army was routed by just such a surprise attack, Washington lost his temper for one the few times in his life.

Perhaps the best thing Washington ever did for Ohio was to pick Anthony Wayne to succeed St. Clair as general. Wayne's thoroughly trained troops were able to defeat an Indian confederation at Fallen Timbers. This victory made Ohio safe for settlement and proved that the new nation could protect its western settlers. It also helped contribute to the decline of British and Spanish influence in the west.

When President Washington left office in 1797, Ohio was well on the way to statehood. The wilderness he had explored as a youth was soon to be added to the country that he had fathered.

CHAPTER II
John Gibson: Unsung Hero

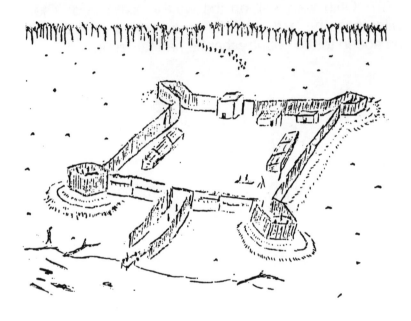

One of the best stories about John Gibson is one that probably isn't true. It concerns the siege of Fort Laurens during the Revolutionary War and Gibson's cleverness in saving the fort. In February of 1779, the garrison at the fort along the Tuscarawas River, the only Revolutionary War fort located in Ohio, was surrounded by Indians led by Simon Girty. The notorious renegade Girty was a sworn enemy of Gibson's and had vowed to personally take his scalp.

Gliding to a Better Place

The starving American troops were reduced to boiling and eating their own moccasins and could not expect any help from the nearby Moravian missions or from far-away Fort Pitt. Yet commanding officer John Gibson allegedly offered the attackers a barrel of flour if they would abandon the siege.

Gibson was not naive enough to believe the Indians would honor their word, but he knew they were also hungry and that they had little patience for siege warfare. Reasoning that the Indians would become discouraged if they felt the garrison had flour to spare, he scraped up enough to fill a barrel, and the attackers took it and left soon afterwards.

There is no verification for this story, but the fact that it was repeated says something about what people thought of John Gibson. And even if this anecdote can't be confirmed, there are plenty of verifiable times where the name of John Gibson shows up at key moments in the strife-filled history of the Ohio Valley. In fact, Gibson played an active role in nearly every major frontier conflict from 1758 to 1814, and his long career parallels Ohio's progress from wilderness to frontier state.

After service in the French and Indian War with the Forbes expedition that took Fort Pitt, Gibson became a fur trader in that area. He was captured during Pontiac's Conspiracy and saved from burning at the stake, possibly by the sister of the Mingo chief Logan, whom he married. Her murder by whites in 1774 started the conflict known as Dunmore's War, and at the end of this campaign it was Gibson who translated the eloquent Logan's Lament.

During the American Revolution, Gibson led a Virginia regiment on the frontier, and by the time of Cornwallis' surrender, was in command at Fort Pitt. After the war, he became a judge and militia general in Pittsburgh. During the Whiskey Rebellion, he sided with the federal government and was one of the prime targets of the mob that invaded the city in 1794.

He began a new career at age 60 when he was appointed Secretary of Indiana Territory. In the absence of Governor William Henry Harrison, he organized much of the new territorial government. He played a crucial role in negotiations with Tecumseh before the Tippecanoe Campaign and again served as acting governor during the War of 1812.

Gibson made many enemies throughout his lengthy career, most notably Simon Girty, but none could question his courage or integrity. He always offered decisive leadership in a crisis and was so fair in his dealing with Indians that his presence was often requested by them during negotiations. He not only worked with seven of the first twelve U.S. Presidents, but with two of their fathers as well. His name often shows up in accounts of the many frontier conflicts during his lifetime, yet he is virtually unknown today. He is a true frontier hero whose story needs to be told.

John Gibson was born in Lancaster, Pennsylvania, on May 23, 1740. His father, George

Gliding to a Better Place

Gibson, had come to Lancaster from Northern Ireland in 1730, and his mother was Elizabeth de Vinez, the Huguenot daughter of a French count. The family also included two daughters and a younger son, George, who was born in 1747.

The elder George Gibson ran the Hickory Tree Tavern on East King Street, and was also engaged in trade with the Indians. Lancaster was on the frontier at this time, and young John Gibson was no doubt exposed to many tales of excitement from the West.

Gibson's mother was more refined and, among other things, taught her children French and Spanish. John found he had quite a facility for languages, which was a skill that proved invaluable on the frontier. Some accounts say Gibson received a classical education, while others say he was self taught. Whatever the source, he was more learned than most of his contemporaries, and his ability with words made leadership a natural role for him.

Gibson left home at age 18 to join General John Forbes' army which was attempting to take Fort Duquesne from the French. Earlier attempts by Washington and Braddock to do this had disastrous results, but Forbes took a cautious approach, and his army was more involved in road-building than combat. His efforts proved successful as the French abandoned the post, which was then renamed Fort Pitt.

It is not known exactly what role Gibson played in this effort, but in 1759 his name first comes up in historical accounts. Here he appears leading troops in battle against some Mingoes led by a chief called Little Eagle. According to one witness, "Captain Gibson, being a very powerful man, entirely severed the head of Little Eagle from his body with one stroke of his sword." The rest of the Indians fled upon seeing this and told their tribesmen that their chief had been killed by a Long Knife, which was supposedly the first time Indians employed that phrase to describe the white men.

At the conclusion of the French and Indian War, Gibson remained in the area and became a fur trader. His principal post was at Logstown, 20 miles downstream from Fort Pitt. He was living here when Pontiac's Conspiracy began.

The Ottawa chief Pontiac was among the first to try to unify the diverse Indian tribes, and his efforts met with remarkable success. With the departure of the French from North America, Pontiac realized that the tribes were on their own against the land-hungry British. He resolved to strike first and in the spring of 1763 an eighteen-tribe alliance overwhelmed every western outpost with the exceptions of Fort Pitt and Detroit. Hundreds on the frontier were taken captive and Gibson was among them.

He was captured along with two companions, one of whom was immediately burned at the stake. The second suffered the same fate after the party returned to their village. Gibson was supposed to be burned, but he was saved by a Mingo woman who adopted him to replace a dead relative. He later took either the sister or sister-in-law of the noted chief Logan as his wife and she bore children by him. He was adopted into the Mingo tribe and given the name Horsehead.

The next year, Colonel Henry Bouquet led an army through the region that liberated all captives, but Gibson retained close ties with the Mingoes. He learned several Indian languages and his familial connections were of great help in his fur trading business.

However, he did not exploit these connections solely for personal gain and, unlike many traders, he always had the respect of the various tribes. The Delaware chief Netawatwes asked to deal with him, saying, "We desire that John Gibson be sent to this Town; as we know him, and he is a Good man." For the rest of his life, Gibson remained a sympathetic friend to peaceful Indians, an attitude that often put him at odds with his white neighbors.

It was the actions of some of these neighbors that gave Gibson a personal tragedy and ended a decade of peaceful prosperity in 1774. On April 30 of that year, Gibson's wife and some of her relatives were murdered by whites across the Ohio River from present-day Steubenville. They had left their village and crossed the river to go to a settlement called Baker's Bottom to get milk for Gibson's infant child. Whites there had gotten some of the Indians drunk, and when a quarrel broke out the entire party was killed, except for Gibson's papoose, who was apparently kept for him.

This barbarous incident touched off a year of border warfare. Logan, who had previously been a staunch friend of the whites, lost all of his relatives in the massacre, and he went on the warpath to seek revenge. The conflict came to be known as Dunmore's War, named for Lord Dunmore, the royal governor of Virginia, who desired war with the Indians to advance the status of himself and his colony.

The land-hungry Virginians were seeking to expand into Kentucky, which was currently hunting grounds for the Shawnee tribe. In addition, since Fort Pitt had been abandoned as a British fort in 1772, both Pennsylvania and Virginia had been claiming the area. Dunmore's agent in Pittsburgh, Dr. John Connolly, had occupied the abandoned fort in the name of Virginia until he was arrested by Arthur St. Clair, a leader of the Pennsylvania faction. By precipitating a border crisis, the Virginians hoped to take more Indian lands and weaken the Pennsylvanians' case.

Dunmore actually had two separate armies ready in the fall of the year. One force under Colonel Andrew Lewis was to proceed down the Kanawha River to the Ohio, while Dunmore himself commanded the larger force, which headed downstream from Pittsburgh. Indians under the Shawnee chief, Cornstalk, attacked Lewis where the Kanawha meets the Ohio, but they were defeated at the Battle of Point Pleasant on October 10, 1774. As the Indians were not able to prevent the two armies from meeting, they were forced to sue for peace after this battle.

One of the most significant aspects of this campaign was that it was the first time an army of Americans had defeated an Indian force without any help from British troops. Some have called Point Pleasant the first battle of the Revolution, and the episode offered many future patriots such as George Rogers Clark and Daniel Morgan an advance taste of military leadership. Gibson served with this army but as a non-combatant. His ability with Indian languages landed him a post as Dunmore's interpreter.

Also serving with Dunmore as scouts were two men named Simon. Simon Butler was a teenager whose brutal beating of a love rival had caused him to change his name and flee to the West--his real name was Simon Kenton. The other Simon was Simon Girty, a Pennsylvanian who had already endured a rough frontier life. Girty had seen his father killed in a drunken brawl and witnessed the burning of his stepfather by Indians, who then held Simon and his brothers captive for the next three years.

After Point Pleasant, the Virginians pushed north of the Ohio towards the Shawnee towns. At Camp Charlotte, near present-day Chillicothe, Gibson was sent ahead to meet with Indian leaders to begin discussing terms. Gibson was engaged in a discussion with Cornstalk when he was approached by Logan, who asked Gibson if he would come with him. They walked to a large elm tree nearby, where Logan tearfully made the following speech:

"I appeal to any white man to say if he ever entered Logan's cabin hungry and he gave him not meat; if he ever came cold and naked and he clothed him not? During the course of the last and bloody war, Logan remained idle in his camp, an advocate for peace. Such was my love for the whites that my countrymen pointed as I passed and said, 'Logan is a friend of the white man.' I had even thought to have lived with you but for the injuries of one man. Colonel Cresap, the last spring, in cold blood and unprovoked, murdered all the relations of Logan, not even sparing my women and children. There runs not a drop of my blood in the veins of any living creature. This called on me for revenge. I have sought it. I have killed many. I have fully glutted my vengeance. For my country I rejoice at the beams of peace; but do not harbor a thought that mine is the joy of fear. Logan never felt fear. He will not turn on his heel to save his life. Who is there to mourn for Logan? Not one."

Logan's Lament is one of the most eloquent speeches in American history, and was later used in the *McGuffey's Reader* series, where schoolchildren across the country read it. Thomas Jefferson used the speech as an example of Native American eloquence in his book *Notes on the State of Virginia*. This proved to be controversial, as, for one thing, the family of Colonel Cresap protested that he was not responsible for the killings. Gibson had actually met

Cresap's party that spring while returning from Fort Pitt, and had in fact been threatened by some of them who felt that Indian traders were even worse than Indians. But while this group did murder other Indians that spring, they apparently were not involved in the massacre of Logan's family.

Some critics accused Jefferson of writing the speech himself, while others claimed that Gibson wrote it. And one eyewitness claimed that Logan addressed the speech to the illiterate Simon Girty, who relayed it to Gibson. In a sworn affidavit over 25 years later, Gibson affirmed that the speech was Logan's and was delivered to him. The speech was no doubt aided by Gibson's able translation and ability with English, and it is also possible that his recognition for his role in this may have been the beginning of Girty's enmity towards him.

After Dunmore had dictated peace terms, Gibson and the Virginians returned to Fort Gower, a crude stockade that William Crawford had erected at the mouth of the Hocking River. Here the troops discussed the possibility of conflict with the mother country. Well aware that trouble was brewing with England, the soldiers drafted the Fort Gower Resolves, a statement that reaffirmed their loyalty but cautioned that the Crown must not trample the liberty of the Americans. Many other groups of colonists were passing similar resolves about this same time, but this frontier version, written a year and a half before the Declaration of Independence, was the only such statement to come from Ohio.

The Ohio Valley frontier in Gibson's lifetime was a place of almost constant conflict, and not just between whites and Indians. The British fought the French and then the Americans, Pennsylvania opposed Virginia, rural Pennsylvanians fought with urban Pennsylvanians. Now as it became apparent that there would be war between colonists and England, old alliances began to shift again.

Lord Dunmore now became more concerned that the Fort Pitt area remain under control of forces loyal to the King, regardless of which colony they were from. Along with his agent Connolly he devised a plan for Tory forces to work with Indians to keep the frontier under British control. Connolly wrote letters to prominent pro-British citizens asking for help in this plot. Because of Gibson's previous close association with Dunmore, Connolly decided it was safe to send a letter to him.

It was not. Gibson's loyalty was with the Americans and he immediately turned this letter over to the local Committee of Correspondence. Connolly was arrested, his plot was foiled, and Dunmore himself had to flee Virginia not long afterwards. This was the first of several times that the revealing of a controversial correspondence played a role in Gibson's life.

In the fall of 1775, a large Indian conference was held at Fort Pitt. The original purpose was to finalize the details of the Camp Charlotte agreements, but the outbreak of the American Revolution changed the focus. Now the Americans used the conference to try to persuade the Indians to remain neutral. An impressive array of chiefs attended, including Guyasuta of the Mingo, Netawatwes and White Eyes of the Delaware, Cornstalk of the Shawnee, Half King of the Wyandots, and from the Ottawa, the son of the great Pontiac. Gibson played a role in the talks that at least slowed down the tribes' drifting towards the British. He was also assigned to supervise the recovery of captives from the various tribes.

Gibson left the frontier not long after this mission. He had joined the Continental Army, and had been named Colonel of a Virginia regiment. He joined Washington's army and served in the New York and Pennsylvania campaigns.

But in the spring of 1778, the desertion of Simon Girty led to Gibson's being sent back to Fort Pitt. Unlike Gibson, Girty had not managed to disassociate himself from the Dunmore/Connelly connection, and his loyalty was questioned to the point that he was even arrested once. On March 30, 1778, he confirmed these suspicions by deserting Fort Pitt for Detroit along with fellow Indian experts Alexander McKee and Matthew Elliott. Girty's brothers James and George followed soon afterwards, which left a dearth of Americans with any Indian experience.

On May 28, 1778, General Washington wrote to the President of Continental Congress that "Colonel John Gibson of the 6th Virginia Regiment, who, from his knowledge of the Western Country and Indian Nations and language, is ordered to repair to Pitsburg (sic)."

Fort Pitt was the center of American activity for the Western Department during the Revolution, and Detroit was the British counterpart. The governor here was Henry Hamilton, called "The Hairbuyer" by the Americans for his alleged policy of paying a bounty for American scalps. An Indian path called the Great Trail connected the two posts, but led through lands mainly occupied by tribes that had aligned themselves with the British.

Detroit was also used as a launching pad for raids on the Kentucky settlements that had been established after Dunmore's War. Raiding parties came down the Miami River to continually harass the Kentuckians. These settlements were considered a part of Virginia and were left to raise their own militia to defend themselves.

The Kentuckians lacked the strength to assault Detroit directly, but under the leadership of George Rogers Clark they came upon the bold plan of conquering isolated British posts in the Illinois country. In the summer of 1778, Clark's men captured the former French forts at Cahokia, Kaskaskia and Vincennes. Clark sent his chief scout Simon Kenton back with word of his success.

That fall Kenton was captured while attempting to steal Indian horses north of the Ohio. He was sentenced to be burned at the stake and was made to run the gauntlet at several villages en route to his execution. Even though he was facing a terrible death, Kenton still noticed how beautiful and fertile the land was and thought it would be a good place to live. He was able to do this 20 years later because of Girty, who met his old friend and was able to intercede on his behalf. Instead of being roasted, Kenton was taken to Detroit, where he was able to escape later.

Meanwhile, at Fort Pitt, the new American commander, General Lachlan McIntosh, also hoped to attack Detroit. McIntosh was a Georgian who was chosen because he was neither a Virginian nor a Pennsylvanian, so he would favor neither state. Also, since he had killed fellow Georgian and signer of the Declaration of Independence, Button Gwinnet, in a duel, it was felt that a change of scenery would help him as well. Unfortunately, McIntosh's blustery authoritarian style soon alienated his allies and he squandered his credibility with the Indians.

In September of 1778, McIntosh called for a meeting at Fort Pitt with all tribes, but attendance was sparse. Most Indians were already aligned with the British, and only the Mingoes, Delaware, and Shawnee, who inhabited land closest to the Americans, would even consider an alliance. However, the Mingo were part of the Six Nation Confederacy that supported England. And all hope of aid from the Shawnee had evaporated the previous fall when Cornstalk had been murdered by American militia when he came to Fort Randolph (at the site of Point Pleasant) under flag of truce.

But the Delaware did attend and on September 19, agreed to a treaty that was the first one negotiated by the new American nation and any tribe. This agreement allowed American forces to pass through Delaware lands and specified that a fort would be built on them. There was also a proposal that the Delaware nation be considered as a possible fourteenth state in the new nation. This was the idea of White Eyes, one of the principal chiefs of the Delaware and a friend of the Americans. Another stipulation of White Eyes was that "it is our particular request that John Gibson may be appointed to have charge of all matters between you and us. We esteem him as one of ourselves; he has always acted as an honest part by us and we are convinced he will make our common good his chief study and not think only how he may get rich."

White Eyes offered to work with the Americans in locating and building a fort, but before this could be accomplished he died suddenly under mysterious circumstances. He was hastily buried and it was reported that he had died of smallpox, which ranked with firearms and fire water as the largest white-introduced destroyers of Indian civilization.

McIntosh left Fort Pitt in November with an unusually large army of 1,200 men. However, the supply train needed to support this force slowed it to the point where the army traveled only about five miles a day. By November 18, they had only reached the junction of the Great Trail and the Tuscarawas River, and it was obvious that Detroit was an unrealistic goal. McIntosh decided to build a fort on the site to protect the Delawares and use it as a launching point for an invasion in the spring. The post was named for Henry Laurens, the President of Continental Congress.

Fort Laurens was located north of -critics said too far north of - a series of Moravian mission towns and the principle Delaware town at Coshocton. The Moravian towns were composed of Delawares who had been converted to Christianity by missionaries led by David Zeisberger. Located in a precarious position, the pacifist Moravians advocated neutrality, but actually provided the Americans with valuable information.

On December 9, the main force withdrew, leaving the unfinished fort and 170 men under the command of Gibson. Although within a few weeks Gibson was able to, in his words, "bid defiance to the enemy", the winter at Fort Laurens was a disaster, with Gibson's calm leadership one of the factors that saved the garrison from annihilation.

The troops were constantly short of provisions and became dependent on the Delawares that they had been sent to protect. And as an isolated post 70 miles from reinforcements, they were a ripe target for Indian harassment. On January 21, a group of Indians under the leadership of Girty intercepted a party that had just left for Fort Pitt. The troops were able to fight their way back to the fort, but a messenger, bearing valuable dispatches that revealed the weakness of the Americans' position, was captured.

When Girty had the letters read to him he became enraged. Gibson had been warned by Zeisberger that Girty was in the area, and in response he had written that if he caught Girty, "I shall trepan him." As a recent deserter, Girty was sensitive to what the American opinion of him was, and hearing that Gibson planned to drain his brain filled him with hatred. After making sure the captured messenger was properly roasted, Girty returned with a few British soldiers and a war party of Mingoes and Wyandots and a personal desire to get Gibson's scalp. The Moravian missionaries warned Gibson that an attack was coming, but the raiders were still able to surprise the fort.

On the morning of February 23, the siege was announced when a wood cutting party of 19 men was ambushed and scalped within sight of the horrified garrison. Then, by using an old trick of parading repeatedly across a distant clearing, the attackers, who had no more men than the fort, were able to further demoralize the Americans by convincing them that they had 847 warriors.

Over the course of the siege, Gibson's men were reduced to boiling and eating their own moccasins and two men died from eating poisonous roots. When one soldier was able to kill a deer, may of the men simply devoured it raw. A relief expedition was sent from Fort Pitt under the command of Major Richard Taylor, the father of Zachary Taylor. This force decided that overland travel was too dangerous and tried to relieve the fort by traveling down the Ohio and up the Muskingum and Tuscarawas. But high water and Indian harassment forced them to turn back well short of their goal.

While Gibson may not have traded the last of his flour to lift the siege, he did know the equally hungry enemy would soon lose patience, and he hoped to outlast them. After being surrounded by hostile Indians for 25 days, the garrison awoke to find the attackers gone. Three days later, a relief pack train showed up from Fort Pitt. The overjoyed soldiers fired their guns in celebration and stampeded the train, resulting in the loss of much of the supplies. After this tragicomic incident the idea of attacking Detroit from Fort Laurens was abandoned, as was the fort itself by the end of summer.

McIntosh was soon replaced by Colonel Daniel Brodhead, who had no use for the fort. Brodhead concentrated his immediate efforts against the Iroquois in New York and Pennsylvania. His decision to give up on Detroit was unfortunate since that post was now vulnerable due to Clark's efforts.

At almost exactly the same time that Girty was failing to capture Gibson at Fort Laurens, Clark succeeded in capturing Hamilton at Vincennes. This opportunity came about when Hamilton left Detroit to recapture Vincennes in December. He then settled in for the winter, confident that there would be no more activity until spring. But Clark took his men cross country through swamps in the dead of winter and surprised Hamilton. Clark also used the trick of parading his men several times through a clearing to magnify their number. Gibson had refused to fall for this trick, but it worked on Hamilton, and the Hairbuyer surrendered to Clark's small force without a fight.

Clark's expedition had been strictly a Virginia sponsored affair, with the mission only known to Governor Patrick Henry and a few other key officials. Henry was succeeded as governor by Thomas Jefferson, who shared Clark's desire to finish the job by capturing Detroit. However, Jefferson preferred a cooperative operation with federal troops and sought a Continental officer to be Clark's second in command.

In January of 1781, Gibson was in Virginia on a recruiting expedition with Baron Von Steuben and he stopped in Richmond to discuss the venture with Jefferson and Clark. They were all able to agree on a plan and Jefferson wrote Benjamin Harrison that "Colonel John Gibson is appointed to go as next in command under General Clark." Harrison was the Speaker of the House of the Virginia legislature and also the father of an eight year old son named William Henry, who would later play a prominent role in Gibson's life. Jefferson also wrote to Washington to get Brodhead to release Gibson and his regiment and to Gibson he wrote, "In the event of General Clark's death or captivity your rank and our confidence in you substitute you as his successor."

Brodhead, however, resisted and claimed he needed all his troops. The Delaware had recently begun to align with the British and in April of 1781 Brodhead launched a preemptive strike against the tribe. He sacked the village of Coshocton and killed several Delaware, including 15 who had already surrendered. After this, he did need all his troops, as the inflamed Delaware joined the British with enthusiasm. Second in command for Clark's venture now fell to militia Colonel Archibald Lochry. In his effort to catch up with Clark's main force on the Ohio, Lochry's entire 100 man detachment was killed or captured at the mouth of the Great Miami River by Indians led by George Girty and the infamous Iroquois leader Joseph Brant.

Simon Girty was not on this raid, but he joined the group a few days later. That night he and Brant got drunk and argued about their respective prowess. When Girty called Brant a liar, Brant struck him with his sword, which gave Girty a scar on his forehead that made him even uglier.

Meanwhile, back at Fort Pitt, the volatile Brodhead had alienated not only any potentially neutral Indians, but also his own troops and the settlers he was supposed to protect. Over 400 citizens signed a petition requesting his removal, and the poorly provisioned soldiers were in a state of near mutiny, which Gibson was in the middle of, as second in command. In September of 1781, Brodhead was removed and Gibson replaced him as temporary commander of the Western Department.

It was a changed situation that Gibson presided over. Any hope of alliance with his friends the Delaware had been shattered by Brodhead's attack. And the other group that he had a rapport with- the Moravian missionaries- was no longer on the scene. That summer the Moravians had warned Wheeling of an impending attack, and when members of the raiding party discovered why their surprise was foiled, the missionaries and their converts were forced to evacuate their towns on the Tuscarawas and go to Detroit to face charges. Though the converts were treated rudely by Girty and spent the winter in discomfort, they were not convicted of anything by British authorities. In the spring of 1782 several of the Indians were permitted to return to their villages to retrieve unharvested corn.

The Moravians had always been in a precarious position as peace lovers caught between warring forces, but had managed to avoid tragedy until now. But Indian raiders had been active over the winter and had used the abandoned Moravian towns as an intermediate base. A party of Pennsylvania militia was looking for the raiders when they came upon the Christian Indians at Gnadenhutten. When a dress belonging to one of the killed settlers was found in the village, the fate of the Moravians was sealed. The militia voted to kill the unresisting Indians, and 90 men, women and children were executed and their bodies burned on March 8, 1782.

Gibson had heard of the militia's movements and tried to stop them, but was too late to prevent the Gnadenhutten Massacre. He railed against the militia, calling them "'the most savage miscreants that ever degraded human nature." For daring to take the part of any Indian over any white man, Gibson was openly threatened on the streets of Pittsburgh.

The Pennsylvanians still wished to mount a punitive campaign into the heart of Indian country. General William Irvine, who had by now been named permanent commander at Fort Pitt, knew his small poorly equipped force would not be able to effectively participate. He offered the services of his army surgeon, Dr. John Knight, but otherwise this was to be a militia operation. In a close election, Colonel William Crawford was chosen commander over David Williamson, who had supervised the Gnadenhutten Massacre.

This army followed the Great Trail to the Sandusky Plains, where they were attacked on June 4, 1782. After a two-day battle, the Americans retreated in disarray and Crawford and Knight were among those captured in the retreat. The Indians would have preferred to capture Williamson, but they were content to wreak their vengeance on Crawford. The colonel appealed to his old comrade Girty for help, and some accounts say Girty did try to get Crawford spared, but the Indians' desire to avenge Gnadenhutten was too strong. Dr. Knight witnessed Crawford's torture, and his account gives the most damaging assessment of Girty, which is the one that survives today.

The unfortunate Crawford was stripped, painted black, beaten, tethered to a post where he had to walk on hot coals, had gunpowder shot point blank all over his body and had his ears cut off. After hours of torment he begged Girty to shoot him and end his misery. But Girty replied that he had no gun and then turned to a companion and laughed. Then, while the Indians were attempting to prolong Crawford's agony, Girty approached Knight and told him to expect the same treatment.

He then tried to engage the horrified Knight in conversation, asking what the Americans thought of him. Knight said that Girty also "expressed a great deal of ill-will for Colonel Gibson and said he was one of his greatest enemies, and more to the same purpose, to all which I paid very little attention." Knight managed to escape a few days later and gave his vivid account of the behavior of the Indians and Girty. This was one of the last battles of the Fort Pitt theater of the war, so there were no more opportunities for Girty to get at his nemesis Gibson. With the war winding down, Gibson left the service on January 1, 1783, and became a private citizen of Pittsburgh.

Gibson continued to play a leadership role in peacetime. He became a common pleas judge for Allegheny County and served as a Major General of the Pennsylvania militia. He was one of the chief negotiators with the Iroquois for purchase of the Erie

Triangle, which gave Pennsylvania a Great Lakes outlet. He also served as a delegate to the state constitutional convention. Although he suffered some financial difficulties stemming from his offering personal credit for supplies during the Revolution, Gibson lived comfortably in a house on Ferry Street during this period.

At several key points in history Gibson stepped out to play an important role, but very little is known about his private life. He was married to a woman named Ann and they had at least one daughter, who cared for Gibson in his last years. One account mentions a son who served in his regiment with him. But Gibson has been so completely hidden in history that it is not even known what he looked like. The one portrait purported to be of him not only is a dark painting of an old man, but also is of questionable authenticity, so no recognized portrait of him exists. All that was passed on was descriptions of his character, nearly all of which mention his courage and honesty.

Gibson was not involved in the Indian Wars of the early 1790's, which was the only border conflict in his long life in which he was not an active participant. However, he did play a role in the early development of the Northwest Territory. In 1788, territorial governor Arthur St. Clair called for treaty negotiations to be held at Fort Harmar, across the Muskingum from the new Marietta settlement. His goal was to certify previous treaty concessions so as to provide for safe settlement of the new lands.

Gibson's experience in treat negotiations made him an obvious choice to attend, and he was one of the Americans to sign the Fort Harmar treaties on January 9, 1789. However, many of the more prominent Indians had boycotted the proceedings, and this treaty failed to prevent a general Indian war.

Gibson did not fight in this war, but his brother George was a participant. The younger Gibson shared his brother's facility with languages, which he put to good use as an officer in the Revolution. In 1776, he led a party of 25 men down the Mississippi to New Orleans, where he negotiated with the Spanish to obtain five tons of much needed gunpowder for Fort Pitt. He got through the Revolution unscathed, but George Gibson was mortally wounded while leading his troops at St. Clair's Defeat in 1791.

Simon Girty was present at St. Clair's Defeat, and he allegedly pointed out the ranking officers among the American dead and wounded so that his comrades could get the most coveted scalps. However, he was once again thwarted in his attempts to get a Gibson scalp, as George's men had carried him from the field to Fort Jefferson, where he died of his wounds on December 14.

John Gibson did play a major role in the Whiskey Rebellion of 1794. This revolt was a response to an excise tax on whiskey that was one of the first tests of the strength of the new federal government. Western Pennsylvanians had no other way of shipping their corn without spoilage than

Gliding to a Better Place

converting it to moonshine whiskey. They deeply resented this new tax and widespread resistance grew stronger.

When a mob burned down the home of the tax inspector and fired on federal troops guarding it, Gibson wrote a letter to the governor denouncing the insurgents. Whiskey Rebels robbing the mail intercepted this letter, and, like Simon Girty before them, used a purloined letter to place Gibson at the top of their enemies list.

An unruly mob was preparing to march on Pittsburgh, and they demanded that Gibson and five others be exiled if the town was to be spared. On July 31, 1794, the frightened townspeople held a meeting at which four of the targeted men readily agreed to leave. Gibson, who presided over the meeting, refused to flee along with one other man, but they were finally convinced that it was the only way to save the city. Gibson apparently walked all the way to Philadelphia, while the insurgents marched into Pittsburgh and left without doing any major damage. Federal troops were called out after this incident, and the revolt was soon suppressed.

Despite alienating some friends and neighbors because of his support of the federal government, Gibson soon resumed his leadership role in Pittsburgh. He might have finished his days in this way, except that duty called him for one more adventure.

In July of 1800, the 60-year-old Gibson was named Secretary of the newly-formed Indiana

Territory. This entity had been separated from the Northwest Territory, as Ohio was approaching qualification for statehood. The remaining 200,000-plus-square miles of the original Northwest Territory was left in the care of Governor-designate William Henry Harrison, Secretary Gibson, and three territorial judges.

Harrison remained in Virginia until January 1801, but Gibson left immediately and arrived at the territorial capital of Vincennes on July 22. He organized the new government and kept records in a cloth bound book he'd brought. He also administered the first census, which showed a population of 5,641 white citizens, including 714 at Vincennes. Gibson is usually listed as acting governor for this period, which makes him the first Governor of Indiana.

When Harrison did arrive, he began to aggressively pursue land acquisition to speed up the process to statehood. This meant extensive negotiations with Indian tribes, an activity in which Gibson could be of great help. He traveled among various tribes in pursuit of this and also assisted in conferences held at Grouseland, Harrison's brick home at Vincennes.

Gibson's work must have proved satisfactory, for he was reappointed by President Jefferson in 1804 and 1808, despite the objections of Treasury Secretary Albert Gallatin, a western Pennsylvanian who held a grudge going back to the Whiskey Rebellion. Harrison had this to say to Jefferson concerning Gibson's renewal: "He is far from being a very expert Secretary, but he is a very honest man, which is much better, and I am persuaded his reappointment would be acceptable to a great majority of the people."

Harrison's land acquisitions helped facilitate the strongest attempt at Indian unification since Pontiac. The Shawnee chief, Tecumseh, and his brother, The Prophet, viewed Harrison's aggressiveness with alarm, especially since they felt some tribes had sold lands they had no claim to. They talked of uniting all tribes so that none could sell land without the consent of all. The charismatic Tecumseh traveled to far off tribes with this message, while his brother stayed at the village of Prophetstown on the Tippecanoe River and provided spiritual justification for the movement.

The Americans feared the growing strength of the movement might be coupled with British backing, and Harrison invited Tecumseh to talk at Vincennes in August, 1810. Tecumseh was not going to suffer the same fate as Cornstalk - he traveled with a retinue of 80 braves.

When he arrived at Grouseland, Tecumseh refused to negotiate inside the mansion, saying that, "the earth is my mother and on her bosom I will recline." He then delivered a lengthy harangue against whites and their betrayals of past treaties and made threats against Indians who had recently ceded lands. As Harrison began to make a rebuttal, he was interrupted by an outraged Tecumseh, who called him a liar and began to denounce him in the Shawnee tongue.

Gibson was fluent in Shawnee and recognized that these fighting words would mean trouble when translated and he called to an aide, "These fellows mean mischief- you'd better bring up the guard." Upon hearing the translation, Harrison drew his sword and Tecumseh his tomahawk, while other armed spectators drew weapons and the unarmed scattered. The tense situation was defused when the guard arrived and talks were suspended for the day.

Gibson's quick thinking averted bloodshed on this occasion, but the inevitable armed conflict took place when Harrison's army marched on Tippecanoe in November of 1811. The ensuing battle was a prelude to the War of 1812 when Tecumseh allied with the British against the Americans.

When war was declared the following June, Harrison pursued a military commission. When he became commander of American forces, this left Gibson as acting governor again. Despite health problems, Gibson endeavored to show a calm leadership on the panic stricken frontier, although he conceded in private that, "My impression is if Harrison is unsuccessful, all is lost in this quarter."

The beginnings of the War of 1812 gave good cause for such despair. On August 16, Detroit had been surrendered without a fight, and the forts at Chicago and Mackinac had fallen soon afterwards. After Detroit had been lost, Simon Girty crossed over

Gliding to a Better Place　　　　　　　　　　　　　　**47**

from Canada and was seen on the streets boasting, "Here's old Simon Girty on American soil again."

In September, Indians attacked Fort Harrison, just 60 miles from Vincennes. The fort was defended by only 15 soldiers under the command of a 26-year-old Captain named Zachary Taylor. This was the son of the man who had tried to rescue Gibson at Fort Laurens over 30 years earlier, and now Acting Governor Gibson returned the favor by dispatching a relief column.

But Taylor, like Gibson before him, held his own and had repulsed the Indians by the time help arrived, The Americans finally had a victory to boost their sagging morale, and Gibson wrote that "the brave defense made by Captain Taylor... is one bright ray amid the gloom of incompetency which has been shown in so many places."

Gibson had begun his military career 54 years earlier, serving with another 26- year-old officer named George Washington, who became the first President, and now he finished by serving with the man who would become the twelfth President. After supervising Taylor's relief, the 72-year-old Gibson relinquished all military command, saying "through the infirmities of old age I feel myself inadequate to the lash of taking the field." Instead, he concentrated his efforts as governor on raising and supplying troops and strengthening frontier defenses.

During the first week of February, 1813, Gibson opened the Fourth Session of the General Assembly of Indiana Territory, and he dutifully sent a copy of his address to President Madison. Although he could be contentious, Gibson was able to work well with the legislature and when they created new counties during this session, they named one in his honor. This session also voted to move the capital to Corydon and by the time they moved there in May, there was also a new governor, as Madison had named Thomas Posey as permanent governor.

Gibson remained as Territorial Secretary through the rest of the war and until Indiana achieved statehood in 1816. He then retired to Pittsburgh at age 76, where he lived at the home of his son-in-law George Wallace. His last years were not pleasant, as blindness and other infirmities plagued him. Blindness also struck his old rival Girty, who died in Canada in 1817. Gibson died on April 16, 1822, and is buried in the Allegheny Cemetery in Pittsburgh.

At the time of his death, the Ohio Valley region where Gibson had spent his life was peaceful and civilized. But during his life it was full of conflict, and few men commanded such universal respect while in the middle of these conflicts. For John Gibson, the story of his life is the story of his times.

CHAPTER III
John Heckewelder: A Bridge Between Worlds

    The frontier was the place where Indian and white societies met, which made it by definition a dangerous place. The chasm that separated the two cultures could be bridged only by special men such as John Heckewelder. And as a Moravian missionary, Heckewelder not only bridged the white and Indian worlds, he also represented a bridge between this world and the next.

From the time he arrived in Ohio as a teenager in 1762, Heckewelder was indefatigable and relentlessly cheerful in his efforts to bring the Gospel to Native Americans. In addition to his work as a minister, he served as schoolteacher to the Indians as well as an author, judge and diplomat. In these many guises he accomplished many Ohio firsts, such as being the first teacher and white bridegroom in what is now Ohio. A man of many illustrious achievements, the modest Heckewelder only wanted to bring together the worlds of the people he served.

John Gottlieb Ernst Heckewelder was born at Bedford, England on March 12, 1743, the eldest of four children born to David and Regina Heckewelder. His father was a native of Moravia in eastern Europe and was a minister in the Moravian Church. This religious sect was the oldest Protestant denomination, as their founder John Hus had been burned as a heretic in 1415, over 100 years before Martin Luther began the Reformation.

The Moravian Church then went underground and continued to suffer religious persecution. In the early 1700's many of them emigrated to Saxony, including the Heckewelders. From there they went to England in 1742. The Moravians stressed education and mission work and soon became interested in the American colonies. They first went to Georgia on the same ship that brought Methodist founder John Wesley, but wound up making their American headquarters in Bethlehem, Pennsylvania, which they founded in 1741.

Heckewelder's parents were involved in mission work in the West Indies, and both of them eventually died of illness there. While they were in the field their children were cared for by church members until they were sent to America in 1754. Young John sailed from London on his 11th birthday and soon was studying in the children's seminary in Bethlehem. He had some lonely times here as most of the other students spoke German while he had been raised in an English-speaking household.

He was apprenticed to a cedar cooper, but had ambitions beyond barrel making. Ever since hearing a sermon on the subject while still a boy in England, he had desired "to serve the Lord in the Missionary field." The Moravians at Bethlehem had almost immediately begun mission work among the American Indians, and had found a receptive audience among the Mohawk and Delaware tribes. In 1762, Heckewelder was offered a chance to assist experienced missionary Christian Frederick Post on a western mission, and he eagerly accepted.

Post had a good reputation among the Indians, and had been instrumental in keeping some tribes from fighting against the English in the French and Indian War. His new mission was to be among the distant Delaware on the Tuscarawas River. He had made initial contacts in 1761, and near the Delaware capital located at present day Bolivar he had built the first cabin by a white man in what is now Ohio. Now he wanted to return with a schoolteacher, and the 19-year-old Heckewelder was thrilled to come along in the role of the first teacher in what is now Ohio.

The pair set out in March and endured snowstorms and floods to go far beyond what was considered the frontier at the time. Arriving at the cabin on the Tuscarawas, they found that Delaware trust in them had lessened in the volatile political climate between the fall of Quebec and Pontiac's Conspiracy. The suspicious Indians, who associated white clearing of land with Indian loss of land, even limited the size of the missionaries' garden, and assigned a brave named Captain Pipe to step off the boundaries.

The eager young missionary also made the discovery that the few instruction books on the Delaware language were worthless and he had to learn the language by immersion. He had had the same problem in Bethlehem when his classmates spoke German, so he was capable of making the same effort, and he soon learned much about the Delaware language and culture. The Indians respected his efforts to learn their tongue, and bestowed upon Heckewelder the name Piselatupe, or Turtle.

Post had previously made a commitment to accompany some Delaware chiefs to treaty talks in Pennsylvania in summer. He promised the Moravian elders he would not leave Heckewelder alone, but as the summer wore on it became apparent that if both missionaries abandoned their post they would not be permitted to return. So Heckewelder offered to remain while Post went east. The teenager encountered problems almost immediately, as both his horse and canoe were "borrowed" and not returned. His garden was regularly raided and he became ill.

One fall day a trader who lived across the river called him over and warned him that he was in danger. Heckewelder spent the night with the trader and the next day discovered that his cabin had been broken into. He now realized it was unsafe to remain and accompanied his trader friend back to Fort Pitt, encountering Post along the way. They returned to Bethlehem in November, and the next spring Pontiac's Conspiracy broke out and missionaries and traders all over the frontier were captured or killed.

While Heckewelder had found it unsafe being one of the few whites in Ohio, the Moravian Indian converts around Bethlehem during the time of Pontiac's Conspiracy found themselves threatened by Pennsylvanians who hated all Indians. The Christian Indians were taken to Philadelphia for their own protection, and when mobs threatened to come after them, the Quakers of Philadelphia armed themselves in preparation. The Moravians were held on a cramped island in the river, where smallpox broke out and over fifty of them died.

After this scare had passed, the Moravian Indians realized it would be safer if they moved west, away from white settlement, and they founded a new mission on the Susquehanna in 1765. Five years later, they felt it necessary to move even farther west to the Beaver River near Pittsburgh. Heckewelder played a role in founding these missions, and often served as a courier between missions. But there was not enough mission work to suit him, and he often had to resume his work as a cedar cooper.

Heckewelder's opportunity came in 1771, when David Zeisberger asked for him as his assistant at the Beaver River mission. Zeisberger was 20 years older than Heckewelder, but he too had begun his missionary career under the tutelage of Post. In 1745, Post and Zeisberger had been jailed for seven weeks in New York City because the pacifists had refused to swear a loyalty oath to the King of England. Now Heckewelder eagerly joined Zeisberger at about the same time that the latter was mulling an invitation to relocate among the Delaware along the Tuscarawas.

This invitation had been extended by Netawatwes, principal chief of the Delaware. Netawatwes, also known as Newcomer, had known the Moravians in Pennsylvania before moving his town west to the present site of Newcomerstown. Having the missionaries among his tribe would help them and advance the chief's status, but Netawatwes also came to be a true friend to the Moravians. After an initial visit where he preached the first Protestant sermon west of the Alleghenies, Zeisberger accepted

Netawatwes' invitation and the first group of converts arrived in the Tuscarawas Valley on May 3, 1772.

They picked a spot south of present day New Philadelphia and named their new village Schoenbrunn, German for "Beautiful Spring". As more converts arrived, they began another town ten miles downstream called Gnadenhutten, or "Huts of Grace". The following spring the Beaver mission was abandoned and moved to the new towns. A large party went overland with the livestock, while Heckewelder led a flotilla of 22 canoes full of converts and supplies on a water route down the Ohio and up the Muskingum and Tuscarawas Rivers.

Here the Moravians had their most successful missions, as for a brief time the Christian Indians were safe from threats of white expansion and far enough upstream to prevent corruption from their fellow tribesmen. The missions grew to nearly 400 members, with over 60 cabins at Schoenbrunn alone. Unlike most Indian villages, these towns were carefully laid out with streets and lots. Schoenbrunn, which Heckewelder called "the largest and handsomest town the Christian Indians had hitherto built", not only featured a church that could and often did hold 300 worshipers, but they even had glass windows and a bell they had painstakingly hauled from Pennsylvania. They also built the first school west of the Alleghenies, where Heckewelder taught nearly 100 students. Zeisberger even published a Delaware language speller and dictionary, so the natives could be instructed in their own language.

The Moravians posted rules of conduct that included Ohio's first prohibition laws. They not only forbade their followers to go to war but even proscribed the purchase of goods known to have been taken in war. The sober and industrious Moravians thrived in their new location and became known for always having food in their larder, unlike many tribes who lived hand to mouth. This led to speculation that some converts were more concerned with the state of their stomachs than their souls.

But an impressed visiting Congregationalist missionary noted that "the Moravians appear to have adopted the best mode of Christianizing the Indians. They go among them without noise or parade and by their friendly behavior conciliate their good will... and gradually instill into the minds of individuals the principles of religion." The Moravians also made an effort to incorporate native language into their ceremonies, as it was noted that in church they prayed in Delaware, preached in English, and sang hymns in German.

The missions would have continued to thrive in blissful isolation had not frontier conflict again caught up with them in the form of the Revolutionary War. The missions were located between the American headquarters at Fort Pitt and the British at Detroit and they soon became involved in the tug of war that convulsed all tribes. The peace loving Moravians had to walk a tightrope between peril on both sides. The danger of their position was summed up in an ominous proclamation by a Delaware war chief who said, "If you pass safely through this war and I see you all alive at the end of it, I will regret not having joined your mission."

The pro-British Indians who raided the American settlements rightfully suspected the Moravians of warning the pioneers of impending attacks. And the American settlers resented that the Moravians fed the Indian war parties and allowed their towns to be used as a launching pad for raids. But to refuse to feed them would have been a serious breach of Indian etiquette, and as Heckewelder observed, "The quickest way to get rid of all warriors is to give them meals victuals, which is all they want, and to refuse them would be folly, as then they would shoot cattle, and destroy the corn in the fields."

While the Moravians were supposed to be neutral, some of the missionaries found it difficult to be impartial. For Heckewelder, it was impossible. He and Zeisberger regularly provided Fort Pitt with information via a fortnightly courier system, and their efforts helped keep the Delaware out of the war at a time when most tribes were going over to the British side. And on one notable occasion, Heckewelder rendered a more active service to the United States.

In 1777, Heckewelder returned to Bethlehem, where he remained for nearly a year. During this time, he was ordained a deacon, and he hoped to marry, but church elders denied him permission after consulting the lot, a method in which the drawing of Scriptural passages was used as a guide in making decisions.

He returned west the next spring, at a time when the church had not heard from the missions in nearly six months. In March of 1778, Heckewelder first met with Henry Laurens, President of Continental Congress before leaving for the frontier. Traveling with one companion, he arrived at Fort Pitt to find the place in turmoil due to the recent desertion of Indian experts Simon Girty, Matthew Elliott and Alexander McKee.

This trio of defectors had passed through the new Delaware capital at Coshocton on their way to Detroit and left many lies in their wake. They alarmed the Delaware with stories that the American army had been decimated and many of the leaders hanged and a motley remnant was now heading west destroying all Indians in their path. The pro-British Delaware under Captain Pipe were using this news to press for an alliance with the British.

The border situation was so unstable that Heckewelder was urged to stay at Fort Pitt, but he insisted on making the trip to the missions. He and his partner rode for three days, just missing a Wyandot war party, and arrived at Gnadenhutten shortly before midnight on April 5. After being informed of the situation at Coshocton, he rested for a few hours and then rode on 30 miles to the Delaware capital, arriving around ten in the morning.

Here he rode into a scene so tense that old Delaware friends would not publicly greet him even though they had not seen him for a year. But when chief White Eyes asked if the stories they heard were true, Heckewelder was able to dispel the tension by refuting the lies.He also had with him newspapers that told of the British surrender at Saratoga, and the Pipe faction was frustrated in their attempts to enter the war. General Edward Hand, the American commander at Fort Pitt, later said of Heckewelder's Ride that "in 1778 the United States were indebted to the active and patriotic zeal of Mr. John Heckewelder, who I firmly believe prevented the immediate commencement of hostilities between the United States and the collective forces of the Shawnee and Delaware nations."

That fall the Delaware signed a treaty with the Americans at Fort Pitt. The Americans were permitted to traverse Delaware lands and promised to build a fort to protect the tribe. But they also expected the tribe to join them in fighting the British.The elderly Netawatwes had died in 1776, and was succeeded as leader of the Delaware nation by White Eyes, who was even more of a friend to the Americans. White Eyes was also more progressive and worldly, having traveled to places such as New Orleans and New York, and he felt that the best hopes for his people lay in joining the U.S. as a 14th state. He had even journeyed to Philadelphia to discuss this plan with Continental Congress. To seal this agreement, White Eyes was given a colonel's commission in the U.S. Army.

Unfortunately, White Eyes died while guiding the American army west. It was reported that he had died of smallpox, but it's possible he was murdered by the Americans he was trying to help. Several years later, the American Indian agent at Fort Pitt arranged for the son of White Eyes to attend Princeton, and in his petition he said that the boy's father had been murdered. Regardless of whether the white man's disease or treachery was the cause, Heckewelder noted that "The death of this great and useful man, was severely lamented by, and a great loss to, the nation; and although his ambitious and political opponent, Captain Pipe, with an air of prophecy uttered 'that the great spirit had probably put him out of the way that the nation be saved'; it was not so considered by the faithful part."

The Americans built Fort Laurens on the Tuscarawas, and Heckewelder visited their commander John Gibson, and regularly offered information and advice. But the presence of U.S. troops offered no protection for the Christian Indians --in fact, on one occasion a convoy of supply-seeking soldiers required a Moravian escort to make it back to their fort safely.

The Moravians continually had to adapt to changes in the war's progress. They had earlier built a mission at Lichtenau, just a few miles from Coshocton, and when they felt they needed the protection offered by the nearby Delaware, they abandoned the other missions and all moved there. At this time, they sent all married missionaries and

Gliding to a Better Place

their families back East, permitting only the bachelors to remain. Later, when the Delaware had begun to align with the British, it was felt to be safer to disassociate from the tribe. At this time, the abandoned missions were reoccupied, and a new mission was founded by Heckewelder at Salem (present-day Port Washington). They felt a little safer now, and not only were the married missionaries permitted to return, but the bachelors were encouraged to marry.

On July 4, 1780 at the chapel at Salem, Heckewelder became the first white bridegroom in what is now Ohio when he married Sara Ohneberg of Bethlehem. Almost exactly nine months later the couple became the parents of the first white girl born in Ohio, a son having born to another missionary couple a few years earlier. But this peaceful interlude proved to be but a false promise.

In the spring of 1781 the Americans attacked the Delaware at Coshocton, executed at least 15 captives and burned the town. Some of the soldiers wanted to attack the mission towns as well, but were prevented by their commander. The Delaware now moved towards Detroit to the Upper Sandusky region, where Captain Pipe and his followers had previously relocated. This left the isolated Moravians as the only Indians left in the Tuscarawas Valley. In August, a large war party under the leadership of the Wyandot chief Half King and Matthew Elliott arrived at the missions. They claimed to be friends and insisted that the Moravians accompany them to rejoin the rest of the Delaware, where they would be safe under British protection. The Moravians did not believe these pretensions of friendship but were in no position to refuse such a "request"'. Still, they hoped at least to negotiate a deal where they would be permitted to stay long enough to harvest their crops.

As their guests were growing increasingly insistent, a portion of Half King's war party returned from Wheeling, where their attack had been unsuccessful. Unfortunately, some of their captives told them that this was because the Moravians had sent warning to them. Now Heckewelder and the missionaries were rudely seized and stripped of their watches and finer clothing while their cabins and possessions were looted. The mission livestock was butchered and left to rot and the entire population forced to accompany their captors. Heckewelder wrote of their treatment that "We had in the whole a very disagreeable Journey, until at last we arrived on the first of October at the old Upper Sandusky Town, where we were left by the Warriors to shift for ourselves." While this exodus was going on, Cornwallis was being bottled up in Yorktown, but the American victory there would come too late to save the Moravian missions.

The missionaries were summoned to Detroit to face charges of treason. Here Heckewelder was astonished to find that their foe Captain Pipe "defended us to his utmost." Possibly he felt bad about how the Wyandot had treated his tribesmen, or maybe he no longer had a grudge against the Moravians since they were no longer in a position to undermine his war efforts. But with the prime witness in effect recanting his testimony, all charges were dropped and the missionaries were treated sympathetically by the British.

However, they still had to stave off starvation at their new village, which they named Captive's Town. After getting through an unusually harsh winter, a large group of converts returned to the Tuscarawas in early March to retrieve unharvested corn. Unfortunately, it had also been a busy winter for Wyandot raiding parties that had used the abandoned mission towns as a way station. By March, loosely organized militia from along the Ohio River were eager to wipe out this nest and to avenge some particularly atrocious tortures that recent war parties had meted out to captives.

On March 7, 1782 a group of 165 Pennsylvania militiamen under the leadership of David Williamson came upon 90 Moravians at Gnadenhutten. By claiming friendship, the militia was easily able to disarm the Indians. But when they found a dress belonging to one of the women who had been captured and killed by the Wyandot raiders, all pretensions of friendship passed and were replaced by a thirst for vengeance. But the militia did not commit a crime of passion so much as they orchestrated a carefully planned mass murder.

They put their captives' fate to a vote, with only 18 soldiers voting to show any mercy. They then locked the Moravian in cabins, where the converts spent their last night praying, singing hymns and preparing for death. The next morning the Christian Indians were taken out in pairs and executed with a cooper's mallet. They were then scalped and the entire village burned. By the end of the day 29 men, 27 women and 34 children had been executed by the Americans.

Two Indians had escaped to warn another party at New Schoenbrunn, and this group brought the tragic news back to the missionaries. Zeisberger was not normally given to despair, but he now lamented in his diary, "nowhere is a place to be found to which we can retire with our Indians and be secure. The world is already too narrow. From the white people, or so called Christians, we can hope for no protection, and among the heathen nations also we have no friends left."

Worse yet, the senseless cycle of violence was still being perpetuated on their behalf. Encouraged by their success, the American militia launched a larger strike aimed at the Upper Sandusky towns. But the militia didn't fare as well against armed heathens and were routed in battle on June 4 and 5. Williamson, who was second in command, escaped, but commander William Crawford was captured and tortured to death in revenge for the Gnadenhutten Massacre.

British officials at Detroit offered the Moravians land near them, and for the next four years what was left of the converts lived at a mission called New

Gnadenhutten. They returned to Ohio in 1786, building a town on the Cuyahoga River. It was here that John Heckewelder left mission service and returned to Bethlehem with his wife, who was seven months pregnant with the couple's third daughter.

Though he never again served as a missionary, Heckewelder remained heavily involved with his church's missionary work. The Gnadenhutten Massacre was such a vile crime it actually brought the Americans to shame, and in 1785 Congress authorized compensatory land grants at the sites of the mission towns. To administer this claim the Moravians organized the Society for the Propagation of the Gospel Among the Heathen, and Heckewelder was named as agent.

Heckewelder was anxious to survey and settle the lands, but the Ohio tribes did not recognize U.S. claims to the land being awarded, so it was necessary to move cautiously. In 1788, Governor Arthur St. Clair of the newly organized Northwest Territory called for treaty negotiations at Fort Harmar at the mouth of the Muskingum. Heckewelder went there to lobby for a speedy resolution. He met with St. Clair and the territorial judges, who were sympathetic but not optimistic. He lodged with Rufus Putnam, the leader of the Ohio Company, which had just settled Marietta across from the fort, and the two became friends. It soon became obvious that it would be unsafe to do any surveying now, but before leaving town, Heckewelder said, "I helped General Putnam lay out a large tree nursery, to which I gave seed on condition I have a share in it when the Brethren started a settlement."

The next year Heckewelder traveled to Pittsburgh, where he hoped to meet U.S. Surveyor General Thomas Hutchins and begin work. Unfortunately, he arrived two hours after Hutchins died and instead of surveying with the man he wound up officiating at his funeral at the request of their mutual friend John Gibson.

As unrest continued and Indian raiding increased, it became apparent that surveying and settlement could not proceed until there was peace. Zeisberger moved his converts to a safer area near Detroit, and then on into Ontario at Fairfield on the Thames River. The tribes of western Ohio experienced great success against the American army in 1790 and 1791, and this forced the Americans to try to negotiate.

In 1792, Heckewelder was asked again to be a bridge between worlds when Secretary of War Henry Knox asked him to assist Putnam in negotiating a treaty with western tribes. Despite a late start the pair traveled downstream from Marietta to Vincennes, Indiana. At Cincinnati they found out from General James Wilkinson that Indians led by Simon Girty had attacked Fort Jefferson on the day that they had announced they would arrive there. This assault gave them pause, especially since two previous peace missions had resulted in the murder of the message bearers. Also at Cincinnati, they were joined by William Wells, a white captive raised by

Indians who had recently defected and now served the whites as an interpreter.

At Vincennes they concluded a treaty with representatives of seven western tribes, and some of the chiefs were to accompany the Americans back to Philadelphia. The swampy marshland in the area had given both Heckewelder and Putnam "bilious fever" and Putnam had to remain behind until he was strong enough to travel. Heckewelder was normally the most mild mannered of men --when he met David Williamson he did not confront him even though he rightfully suspected that this was the same man responsible for the murder of his people at Gnadenhutten. But he could be forceful. When the army officer charged with protecting the chiefs on the return trip persisted in supplying them with alcohol, Heckewelder declared "I told him positively that if he continued like this, I would not travel with him another step and would lay complaint against him in the proper place.This frightened him. He begged my pardon and promised he would follow my advice in every particular."

Heckewelder developed a rapport with nearly all Indians he met, and this group was no exception. When the chiefs left Heckewelder at Marietta to go on to Congress, the troops at Fort Harmar fired a seven gun salute. The chiefs then insisted on "one more shot for our friend" , and an additional gun was fired to salute Heckewelder. However, the Indian insistence on the Ohio River as a boundary resulted in Congressional rejection of the treaty negotiated by Heckewelder and Putnam.

The next year the Americans tried to send representatives to deal directly at the huge Indian conference held at the confluence of the Auglaize and Maumee Rivers. Heckewelder was again named as an assistant to the American Peace Commissioners. His group traveled from Philadelphia to New York and then on to Niagara Falls, where they had to wait until British officials could guarantee their safe passage.

They were permitted to go on to Detroit, but never got near the conference at The Glaize. The British claimed neutrality but were trying to keep the Americans out of the peace process. After receiving insulting messages and demands that the Ohio River be the permanent boundary, the commissioners realized the Indians welcomed war, and they returned empty handed. Heckewelder made a side trip to visit Zeisberger at Fairfield and then returned by way of the St. Lawrence to Montreal and then down Lake Champlain.

With the issue now left up to the warriors, Anthony Wayne had greater success than the peace commissioners, defeating the Indians soundly at Fallen Timbers. The Greenville Treaty of 1795 opened most of Ohio for settlement, and the next year Congress reaffirmed the Moravian grants. In May of 1797 Heckewelder was finally able to return to Gnadenhutten, 15 years after the Massacre.

He was pleased to find the apple and peach trees still thrived at the site, but the overgrown village still contained the bones of the butchered brethren. After a brief clean up, he hurried to Marietta, where Putnam had recently been named as Surveyor General. Putnam and his son returned with Heckewelder and did the survey that was a necessary prerequisite for settlement. While doing this and staying in a cabin they had built, Heckewelder presented Putnam with a book detailing the history of the Moravian missions. Putnam read it and marveled that he was "reading it in the very country and indeed on the very spot where an Indian congregation had lived and where the captivity had taken place."

The following spring Heckewelder traveled to Fairfield. After conferring with Zeisberger, they decided that Heckewelder would return to Ohio and prepare the sites and Zeisberger would follow with a group of converts later. Heckewelder's group moved into their first home in Gnadenhutten in September and they had dwellings ready for Zeisberger's contingent when they arrived in November.

The Moravians had great plans for returning to the scene of their most successful missions. Zeisberger was to live with the converts at Goshen, the site near the old Schoenbrunn mission, and use this as a training ground for new missionaries. Heckewelder was to use the downstream Gnadenhutten site as a bridge between the mission and white worlds. Gnadenhutten was to be settled by white Moravians from Bethlehem who would provide trade outlet for Indian goods while insulating the converts from corrupting white influence. Heckewelder was very enthused about this project and spoke of Gnadenhutten becoming a new Bethlehem.

For the first few years Heckewelder spent the summers in Gnadenhutten before wintering in Bethlehem, but in 1801 he moved his family to Ohio year round. It was the third time in 40 years that he had moved to the Tuscarawas Valley, and he was still more than 50 miles away from any other white settlement. But the new town thrived and by 1801 offered the services of a grist mill, sawmill, blacksmith shop and trading post.

Heckewelder also became a prominent local citizen, serving as postmaster and justice of the peace. After statehood was attained and Tuscarawas County formed, he was named the county's first Judge of the Court of Common Pleas. His reputation for integrity spread far enough that the state legislature put him on a panel responsible for the selection of new county seats. He also found time to pursue his hobbies and even published a meteorological diary in the *Philadelphia Medical and Physical Journal.*

But the influx of white settlers made things difficult for the converts at Goshen. Most pioneers were suspicious of the Indians and some merchants from nearby New Philadelphia persisted in selling alcohol to the Indians and then taking advantage of them. The mission lasted until 1821 but foundered after the death of Zeisberger. The venerable missionary "went to the Savior" on November 17, 1808, having lived the last 62 of his 87 years among the Indians. Heckewelder was with him on the day he died.

In 1810, the 67-year-old Heckewelder retired to Bethlehem. He made one final visit to Ohio in 1813, but otherwise his travels, which he estimated covered over 30,000 miles, were over. In the same year as his last trip was the last battle in which Ohioans fought Indians. This battle, where Tecumseh was killed, was fought at the site of the Moravians' Canadian mission at Fairfield. The village was destroyed in the battle, so the tragic results of the Moravians' efforts to avoid war never changed.

After returning to Pennsylvania, Heckewelder began a new career as an author. His wife died in 1815, so writing helped fill the void of his widower years. At the behest of the American Philosophical Society, he drew upon his Indian knowledge to publish *History, Manners and Customs of the Indian Nations Who Once Inhabited Pennsylvania and Neighboring States* in 1818. This book proved to be popular and was soon translated into German and French. Heckewelder was criticized by some scholars for naive acceptance of Indian legends, but his first hand experience make the work a valuable ethnographic study at a time when little other information was available.

He followed this book with *A Narrative of the Mission of the United Brethren among the Delaware and Mohegan Indians* in 1820. He also published some of his travel diaries, which offer excellent descriptions of Ohio towns in their infancy. A book on the Delaware language was published posthumously.

When Heckewelder died on January 31, 1823, one obituary said, "with Mr. Heckewelder has died more critical and accurate knowledge of the Indians' customs, history and language than is left behind him.....When we consider his untiring benevolence, his patience in enduring privations and fatigue, the motives that actuated him...he appears before us as an extraordinary man. He deserves to rank among the wisest and best of his generation, and as one of the benefactors of mankind."

CHAPTER IV
Anne Bailey: Lady of Legend

Life was hard for men on the early frontier. But it was even harder for the women who shared all the hardships yet got little recognition for their efforts. Of course, they were sometimes not included at all-- when the original 48 settlers landed at Marietta on April 7, 1788 they were all male.

The presence of a woman changed the nature of any wilderness expedition. When Lewis and Clark went west a few years later, the Shoshone guide Sacagawea was the only woman among the thirty some explorers. But her presence may have saved the expedition since it indicated to local tribes that the white intruders were not a war party.

In the Ohio Valley, however, the presence of women indicated to the Natives that the whites had come to stay, so women were often singled out on raids on frontier homes. So once women arrived at their new homes, they were given more than their fair share of both the drudgery and terror that characterized frontier life. And since their domestic chores virtually bound them to their homesteads, the only way a woman could distinguish herself in any way was to behave like a man.

This is what Anne Bailey did. After 1774, this short, stout Englishwoman wore buckskin under her petticoat and transformed herself into one of the best scouts on the frontier. She served as an unofficial recruiter and messenger during the American Revolution and saved the town of Charleston, West Virginia by making a 200-mile solo ride through hostile territory. She also played a role in the settlement and survival of the French settlement at Gallipolis, which is where she spent her final years. As a hard riding, hard swearing, whiskey drinking, tobacco chewing frontier scout, Anne Hennis Trotter Bailey's exploits approached the legendary.

As often is the case with legends, there is some confusion about the actual facts. But it is now clear that Anne was born in Liverpool, England in 1742 as Anne Hennis. She was named in honor of England's Queen Anne, whom her father had served at the Battle of Blenheim in 1704. At the age of five, young Anne accompanied her parents on a trip to London, where they witnessed the execution of a Scottish Jacobite for treason. The event made an impression on the girl as she told the story in later years. It also served as reference point to ascertain just when she was born.

The Hennis family was apparently middle class as Anne received some education. Unfortunately, they must not have been healthy, as both parents died while Anne was in her teens. Left with little money and apparently no family, the young girl emigrated to America in 1761.

It was said she paid for her passage by becoming an indentured servant. After working off her debt she moved west and at age 19 settled in Augusta County, Virginia near the present day town of Staunton.

Here she met Richard Trotter, an experienced backwoodsman who had served on General Braddock's ill-fated campaign in 1755. The two fell in love and were married in 1765. Two years later, Anne gave birth to a son they named William. Staunton was on the edge of civilization at that time, but Anne adjusted well to the rigors of frontier life.

But all hopes for normal domestic life were shattered for Anne during Lord Dunmore's war in 1774. During this period of violence, the Virginia militia was called out to fight the Shawnee, and an experienced veteran like Richard Trotter was expected to do his part. He was killed at the climactic Battle of Point Pleasant on the banks on the Ohio on

October 10, 1774, leaving behind a 32-year-old widow and a seven year old son. This tragedy had a great effect on Anne, and she was never the same afterwards.

Upon hearing this news, she entrusted the care of her child to a neighbor and started to live a nomadic life. She began to dress as a man, complete with hunting knife and rifle, as she lived out among the elements. Maybe it was what she wanted all along, or maybe grief drove her mad, but for the rest of her life, Anne wandered from one outpost to another without a fixed home.

During the American Revolution, Anne served as a sort of unofficial recruiting agent, exhorting able bodied men at all stations to join in fighting the British and their Indian allies. Her wandering nature also made her useful as a messenger, since this was a time when most people feared to travel, especially alone. So Anne took dispatches between forts and gradually expanded her travels.

She wound up spending a lot of time in Fort Savannah, the present site of Lewisburg, West Virginia. Here she met a soldier named John Bailey who was not put off by her peculiar ways. The two fell in love and were married in Lewisburg on November 3, 1785 in a ceremony performed by the first Presbyterian minister west of the Alleghenies.

But marriage did not change the lifestyle of Anne Bailey. She continued to travel alone between settlements, although her husband's transfer to Fort

Lee (present day Charleston, W. Va.) in 1788 did mean that she spent a little more time at that place.

It was there in 1791 that Anne was called upon to perform her greatest feat of heroism. Although independence from Britain had been won, British forts on the Great Lakes were still encouraging Indian raids on frontier settlements. Recent victories over American forces had encouraged the Indians further and by 1791 no outpost felt safe.

At Fort Lee, a large group of Indians was seen in the area, which led the garrison to believe an attack was imminent. But much to their horror, they discovered that the fort's supply of gunpowder was woefully inadequate to withstand a siege. Colonel Clendenin, the post commander, asked for a volunteer to ride 100 miles through Indian infested territory to Lewisburg to get powder.

No man in the garrison was willing to undertake this mission. But one 49-year-old woman was. Anne Bailey stepped forward and was given the fastest horse in the fort, which she rode non-stop to Lewisburg. The commander there gave her another horse to carry the powder and offered an escort back to Charleston. Anne declined the offer, preferring as always to travel alone.

She returned in time to save the fort and receive a heroine's welcome. In describing her reception in later years, Anne reported that "the general said 'you're a brave soldier, Anne' and told the man to give me a dram. God, I love a dram!"

But whiskey was not Anne's only reward. The grateful garrison let her keep the horse she had used on her ride. She named the big black horse with the white feet "Liverpool", after her hometown. The two were nearly inseparable for years afterwards.

It was at about this same time that Anne became involved with the new settlement at Gallipolis on the north shore of the Ohio River. The western terminus of her previous travels had been Fort Randolph, where the Kanawha meets the Ohio. This was the site of the Point Pleasant battlefield where Anne was first widowed, and was the most isolated of the Virginia posts. But the opening of the Northwest Territory to settlement changed that and the opposite shore from Fort Randolph was soon host to 500 French settlers.

The opening of the Ohio country led to widespread speculation in land. One of the leading speculators was Congressional lobbyist William Duer, who helped organize the Scioto Company. This private company aimed for a foreign market and sent Joel Barlow to Paris as agent.

Barlow was a poet, not a real estate salesman, and he had little success until he hired an Englishman named William Playfair to help sales. The misnamed Playfair proceeded to describe elaborately an Eden in the wilderness. Not only was this a beautiful land with a temperate climate and wonderful hunting and fishing, but there were trees that produced sugar and plants that grew candles. These last exaggerations referred to maple trees and

Gliding to a Better Place

cat-tails, and Playfair used plenty of other real estate talk to describe the swampy land that was miles from civilization.

These wild claims proved to be quite successful and sales were brisk. After all, this was Paris just after the fall of the Bastille and there was a lot of justifiable uncertainty about the future. Unfortunately, the buyers tended to be middle class artisans with money but few practical skills. There were plenty of wig makers, but not enough woodcutters.

In February of 1790, nearly 500 of these Parisians sailed to America. After a rough voyage, they arrived in Virginia and discovered that they had been duped-- the Scioto Company didn't own clear title to the land sold to them. Many Americans were sympathetic to their plight, however, and went to great lengths to assist our former Revolutionary allies. Rufus Putnam of the Ohio Company hired workers to build 80 cabins on the site so that the immigrants could move right in. The French 500 traveled overland to Pittsburgh, then down the Ohio by boat, landing on October 17, 1790.

They named their settlement Gallipolis (The City of the Gauls), and celebrated by immediately throwing a ball. This became a semi- weekly tradition that continued no matter how grim the outlook was in the community. And it was grim indeed for a while. The new settlers floundered as they lacked the most basic skills. Some of them were struck by the trees they were trying to cut down. This was also during the peak of paranoia about Indian attacks, and malaria was a constant threat along the riverbanks. It was often a miserable existence for these unlikely pioneers, but they persevered and eventually prospered.

Within a few years, the locals were producing a pretty good peach brandy, and several of the craftsmen had found niches. The Moravian missionary John Heckewelder visited town with Putnam in 1792 and complimented the "skilled mechanics and beautiful gardens" he found. He had particular praise for the watchmaker, stone mason and glass blower whose works he saw.

These frontier artisans were able to ply their trades partly because they were supplied with materials by Anne Bailey. Her peripatetic nature was of benefit during wartime as a scout and messenger, but her skills also translated well to peacetime. As one who crossed the mountains frequently, Anne was able to handle specialized supplies or requests for items not normally found on the frontier. As a regular bearer of mail, coffee, medicine, powder, equipment and small hardware, Anne soon became "a great favorite" and frequent visitor to the cabins at Gallipolis.

Anne's reputation for honesty and reliability got her plenty of assignments at all of her regular stopping points. She even branched out into hauling livestock across the mountains. She became one of the drovers who brought pigs and cattle to the west, and on one occasion brought the first tame geese to the Kanawha Valley.

Her client for this venture was Colonel Clendenin from Fort Lee, who was known to very particular in his business dealings. In this case, he specified that he had to be supplied with exactly 20 geese. One goose died along the way, and when the Colonel counted only 19 upon delivery he announced he would not pay. Upon hearing this, Anne drew the dead goose from her saddlebag, threw it at Clendenin's feet and said "There's your twenty." She was promptly paid.

Stories like this about Anne abounded, and she may have contributed to them herself, as she was known to be a lively raconteur. The stories are all that is left after the physical evidence is gone. And even Anne Bailey's Cave, where she would spend the night on trips between Charleston and Point Pleasant is gone now, a victim of excavation.

In her later years, Anne often regaled listeners with exciting accounts of her escapades. She told about how she almost froze one night but was saved by Liverpool's breath. It was so cold she burrowed into a hollow tree for the night and held her horse close so that Liverpool's breathing on her was a source of warmth.

Even more dangerous was trying to avoid hostile Indians. Anne's closest call came when a raiding party overtook her and she was forced to dismount and escape on foot. She hid inside a sycamore log and escaped detection, even though some of the natives rested on the very log where she was hiding. Not content to have escaped with her life, she followed on foot and that night sneaked into camp and untied Liverpool. Then with a whoop that awakened her pursuers, she rode off and got away.

The Indians gradually came to leave the "white squaw of the Kanawha" alone. They recognized her unusual behavior as being possibly insane, and to them anyone so touched by the Great Spirit should be left alone. So Anne was generally accorded free passage, which only increased her legendary status. Not that Anne would have been afraid anyway. Concerning her adventures, she said, "I know I could only be killed once, and I had to die sometime."

Anne's reputation grew to where nobody was really sure what the facts were. People weren't even sure how old she was. The story of her visit to London at age five somehow got garbled, although Anne herself may have had a hand in the confusion.

It came to be believed that that she was there to witness the coronation of her namesake Queen Anne, which would have put her birth around 1700. Arithmetic was not a pioneer strong suit, for if anyone had done the math they would realize how unlikely this was. For this would mean that she gave birth to her only child at age 67 and that her daring ride to save Fort Lee was made while in her 90's. Yet for years after her death, many seriously believed she was born in 1700.

Anne's wandering ways also served to increase her legend. The pioneers may have crossed oceans and mountains to get to their new homes but once they got there they rarely left plow and hearth. Someone like Anne, who could show up at any time and who had such flamboyant characteristics, was sure to be talked about.

In this way, Anne was comparable to fellow frontier legend Lewis Wetzel (1764-1808), the renowned Indian fighter based around Wheeling. Along with the Zane family, Wetzel was among the first settlers in an area that experienced particularly violent border warfare. Wetzel's family was like may in regards to losing members to Indian raids, but Lewis surpassed all in his desire for vengeance. His simple goal was to kill as many natives as possible.

In time of war, the settlers had no better friend than Lewis Wetzel. His shooting and tracking skills were unsurpassed, and he became known for crossing the Ohio to rescue white captives who had been taken in raids. He was an intense, quiet man who grew his hair to his knees to taunt the Indians by challenging them to try and take his distinctive scalp.

But those who accepted the challenge had to beware. Wetzel had taught himself to do the 23 separate steps involved in reloading a flintlock rifle while on a dead run. This meant he could reload while fleeing and then turn and fire on Indians who outnumbered him in gradually decreasing numbers.

But unlike Anne Bailey, Wetzel's skills did not translate well into peacetime. Killing Indians was not an advantage then, since it jeopardized the peace, but Wetzel could not help himself. During the Revolution, either Wetzel or his brother Martin (again, it's hard to be sure when talking about legends) came up to a negotiating chief and tomahawked him while he was speaking to the American commander. He escaped punishment as his fellow soldiers saw nothing wrong with this, but federal officials often felt differently.

In 1789, Wetzel was in Marietta while Governor St. Clair was negotiating with Indians for land concessions at Fort Harmar, just across the Muskingum. Wetzel waylaid and shot a visiting chief, which threatened the treaty talks. For this open murder he was arrested by General Harmar but he escaped with the assistance of sympathetic pioneers.

Among his fellow settlers, Wetzel was a hero and the federal government was part of the problem. The Indians were the enemy and the government was not only failing to protect settlers from raids, but they were evicting white squatters from their homes. To prosecute a white man for killing an Indian seemed the height of governmental folly. So Wetzel escaped punishment for this murder, but eventually had to move on to more fertile killing fields. Ironically, this illiterate scout did go to prison later, but for the unlikely crime of counterfeiting. This was after he had relocated to the Natchez area, where he died around 1808. He was buried there but later reinterred in West Virginia, where he remains a folk hero.

Gliding to a Better Place

Anne Bailey actually is more comparable to Johnny Appleseed as legends go. Both were eccentric, strangely dressed wanderers who were nonetheless respected for the work they did. And both went against traditional gender roles. The barefoot pacifist vegetarian Johnny Appleseed certainly wasn't the prototype of the frontier male, and no woman ever acted more like a man than Anne Bailey.

In both their cases, sympathetic pioneers sought to explain their behavior by invoking the insanity defense on their behalf. They said that Johnny had been kicked in the head by a horse as a lad and that Anne had had become unhinged by grief when her first husband was killed. In fact she was often called "Mad Anne"-- but never to her face. It is also interesting to note that there is no record of her having killed anyone, although it can be assumed she would be quite capable of it should circumstances require it.

Anne gradually moved her activities to the Gallipolis/Point Pleasant area as she grew older. She had remained close to her son even though she hadn't raised him, and William Trotter eventually settled near the mouth of the Kanawha at Point Pleasant. In 1800, he canoed across the Ohio with Mary Ann Cooper and they were married in Gallipolis, allegedly the first Virginians to be married in the French town. John Bailey died in 1802, leaving Anne widowed again, and she started to spend more time at her son's, although she still traveled widely.

In fact, as late as 1817, Anne was still wandering. In that year the 75-year-old woman traveled alone on foot (having outlived her horse as well as her husbands) for a last visit to Charleston-- a distance of 120 miles round trip.

But by this time, she was primarily living with her son and his family, and when William bought land across the Ohio the next year, Anne moved with them. At first, she didn't want to leave the river for an inland farm, so she built herself a shelter overlooking the Ohio. Later, she moved to William's farm nine miles from Gallipolis, although she still maintained separate quarters that she built herself.

Although she was now living under a roof near her ten grandchildren, Anne did not become completely domestic. She still regularly walked or canoed to town with her rifle and enjoyed her status as a local character. She continued to drink, chew tobacco and regale the locals with tales of her escapades. Apparently she never lost her Liverpudlian accent, which partly consisted of adding the letter "h" to words starting with vowels. So she would tell about the time she "shot a howl on a helm tree across from the mouth of the Helk River."

In late November of 1825, Anne said she didn't feel well and canceled a planned trip to Gallipolis. It was unusual for the 83-year-old to complain, so her daughter-in-law sent her two youngest daughters to spend the night at Anne's shelter. The eight and six year old girls were excited to spend the night at Grandma's, but sometime during the night of November 22, Anne died peacefully in her sleep. After a rough life spent mainly sleeping alone outdoors, this rugged frontierswoman died in bed with her two young granddaughters on a sleep over.

Anne was buried at Bailey Chapel in Gallipolis, but her body was later moved to Tu-Endi-Wei Park on the site of the Battle of Point Pleasant. A small monument marks the final resting place of one of the most legendary women of the American frontier.

CHAPTER V
James Wilkinson: They Weren't all Heroes

General Anthony Wayne faced a formidable task as he was preparing the U.S. Army to fight the Indian confederation in the Ohio country in 1794. To begin with, he was given an ill-equipped, poorly-trained and unsupported army that was still smarting from recent disastrous defeats. And he was facing an overconfident and unusually unified band of Indians that was eager to fight. And the people on the frontier that he was supposed to be protecting had no respect for his force and openly doubted their federal government's capacity to do the job.

In addition to all this, Wayne had one more lesser-known obstacle to overcome: he had to deal with a second-in-command whose goal was to undermine and replace him. General James Wilkinson ridiculed his commander in private correspondence, sabotaged his supplies and intrigued with political cronies to have Wayne removed. And this was only a part of his double-dealing. Amazingly, he performed all these disloyal activities while on the payroll of the Spanish government as Agent Number 13.

This was just a part of the fascinating career of Wilkinson, who was called by historian Frederick Jackson Turner "the most consummate artist in treason that the nation ever possessed." Over a military career spanning nearly 40 years, Wilkinson's greed and love of intrigue led him to betray a virtual "who's who" of the founding fathers, but his treason was never uncovered until long after his death. His scheming and duplicitous life serve as a reminder that it wasn't just heroes who settled the frontier.

James Wilkinson was born in 1757 in eastern Maryland of a land holding family. His father died when young James was nine, but he left enough property to support the family and provide for a decent education. It was decided that the boy should study medicine and at age 15, he was sent to Philadelphia to do so. He loved the city and in particular became enamored of military pomp when he saw British troops performing drills there. At 17, he returned home qualified to perform minor medical duties, and he also joined a local rifle company.

When the American Revolution broke out in April of 1775, Wilkinson's fondness for military matters made it obvious that he would get involved. Still, he didn't join the American cause until August, but then James Wilkinson never did anything quickly. With his educated background and an ability to impress the right people, Wilkinson was made an officer despite his being in his teens. He was as hard-working as he was ambitious, and he acquired a reputation as a disciplinarian, something that was needed in the rag tag Continental Army.

For his baptism of fire, Wilkinson was sent on the American invasion of Canada. Also on this expedition were Benedict Arnold and Aaron Burr, two future scoundrels whose treasonous exploits would not match Wilkinson's. All three served well this time, but the campaign was a failure and the Americans had to retreat to New York.

In 1777, Continental Congress sent General Horatio Gates to command this northern army, and Wilkinson found a star to whom he could hitch his wagon. Gates had been an officer in the British Army, but found that his advancement possibilities were limited because his family was of the servant class. He offered his services to the Americans, who were pleased to have an experienced soldier and gave him a high rank. But Gates' sensitivity about his humble origins made him highly susceptible to flatterers. And Wilkinson, it seems, was an ardent sycophant. At the beginning of the Saratoga campaign, Wilkinson was serving under General Arthur St. Clair at Fort Ticonderoga, but not long after Gates' arrival, he was transferred to his staff and given a promotion.

The British plan was to isolate New England from the rest of the colonies with a three-pronged assault that would meet around Albany. However, only the army that came down Lake Champlain under General John Burgoyne followed the plan. This isolated force had a long line of supply and moved slowly through the woods. They were ultimately forced to surrender after meeting stiff American resistance, and the victory persuaded France to join the American cause.

On the eve of the first battle at Saratoga, John Hardin, an officer in Daniel Morgan's crack regiment, captured some British troops who gave them valuable information. Hardin turned this over to Wilkinson who passed it on after writing a report stating that he and Hardin were on reconnaissance together and Wilkinson had made the capture. This theft of credit was pretty much the extent of Wilkinson's active participation in the campaign.

But he played a much larger role in the surrender process, acting as Gates' chief agent. This involved such 18th century military protocols as blindfolded trips to enemy headquarters, ferrying counter offers between camps, and introducing the principals to each other. It was heady stuff for the 20-

year-old Wilkinson. A mere five years after first being impressed by British drill teams, and there he was supervising the surrender of a 7,000-man British and Hessian army.

An even greater reward came when Gates selected Wilkinson to take the news of his victory to Continental Congress. Among the dispatches Wilkinson carried was a recommendation that he be promoted to general. Wilkinson turned this trip into a grand tour, taking two weeks to make a journey that the mail service could do in half the time. He stopped at the headquarters of General Stirling, where he got drunk with some fellow staff officers, and he stopped off to see a girl he was courting. By the time he finally arrived, news had already been received, and one congressman cracked that Wilkinson's reward should be a pair of spurs. But Gates was the hero of the hour and could hardly be denied any request, so Wilkinson became a General in the Continental Army while just barely out of his teens.

Almost immediately, he became involved in controversy. There was movement among some officers to replace Washington with Gates. General Thomas Conway was at the center of this cabal, and Gates gave it clandestine support. But Washington got wind of the plot and warned Gates of disloyal officers. Gates obviously wanted to know who had leaked word, and Wilkinson reminded him that Washington's aide Alexander Hamilton had been in camp recently and suggested he might be the culprit.

When Gates wrote Washington to tell him not to believe Hamilton, Washington replied that his information came from Wilkinson.

Apparently Wilkinson had told Stirling's officers over drinks about the plot in hopes of recruiting them, and the loyal Stirling had informed Washington. Gates was furious about being compromised, and he and Wilkinson nearly fought a duel before reconciling at the last moment. The cabal collapsed, and instead of replacing Washington, Gates was sent south where his command suffered the worst American defeat of the war. As for Wilkinson, Washington mistakenly believed that he had revealed the plot out of loyalty, so he actually benefited from his drunken betrayal. This kind of luck would hold for Wilkinson over the years in future, and more calculated, betrayals.

He stayed away from the battlefield for the rest of the war, serving as the army's clothier general. This gave him opportunity to learn firsthand about the graft and kickbacks that were prevalent in military supplies. In 1778, Wilkinson married Ann Biddle, the girl he had stopped to see after Saratoga. She was from a prominent Philadelphia patriot family, and despite all the vices that Wilkinson eagerly embraced, he was apparently true to his wife. The couple settled in Pennsylvania after the Revolution, where Wilkinson dabbled in farming and politics. But he was too ambitious and adventurous to live such a staid life, and he saw that the West was where opportunity lay. So in 1784, the Wilkinsons moved to Kentucky.

At age 27, Wilkinson was not a physically imposing presence. He was short and already acquiring an ample girth that would grow larger over the years. He was vain and impressed by fancy uniforms, according to a French diplomat who knew him. But he was impressive in other ways. His educated background was apparent in his conversation and even more so in his florid prose style. These alone would hardly have impressed the Kentucky frontiersmen had he not also possessed a great personal charm and an ability to make a strong first impression. With these attributes he set out to become a leader in Kentucky.

It was a situation that cried out for leadership. Kentucky was considered a part of Virginia, which was a state in the still-forming United States. As the largest concentration of settlers west of the Appalachians, Kentucky had a unique set of problems, represented by Indians and the Spanish. Indian warfare had been bitter throughout the Revolution and tribes from north of the Ohio were still a constant threat. To the south, Spanish control of the Mississippi at New Orleans throttled western commerce. Whatever form of government Kentucky was to have would have to deal with these two problems.

Many felt that Kentucky would be best served by joining the union as a new state. But there was not yet a U.S. Constitution that provided for this.

Neither the loosely aligned federal government nor the state of Virginia could provide protection from Indian raids, and eastern leaders were seen as unsympathetic to western needs. Kentuckians were outraged, for example, when federal treaty negotiators proposed bargaining away Mississippi navigation rights in order to gain concessions for eastern shipping.

Soon there was talk of forming a separate country that could negotiate its own treaty with Spain. This was not seen as treasonous, since the same principles that caused the colonies to separate from England could be used for Kentucky to leave the United States. Kentuckians saw themselves as free agents willing to bargain with whomever could best alleviate their problems. The Spanish welcomed such a movement, as they were afraid the ever increasing numbers of Americans pouring across the mountains would overwhelm their lands and take them by force.

Wilkinson soon became involved in the statehood movement and attended a series of conventions that were held to discuss the issue. But he didn't make his bold move until the summer of 1787, when he floated a cargo of goods down to New Orleans. Anyone doing this risked arrest by Spanish authorities and possible confiscation of their cargo. But Wilkinson had no intentions on being so passive.

He allowed himself to be arrested, but asked to be presented to leading authorities. The Spanish were impressed to meet a well-spoken American who had been a General in the Revolution, and they were even more impressed at what he had to say. Wilkinson proposed a deal where he would do his best to separate Kentucky from the United States and align it with Spain and he would keep them informed of all anti-Spanish activity in the West. In return, he desired a Spanish pension and unfettered trading rights on the Mississippi. The Spaniards were eager to agree to this and on August 22, 1787, Wilkinson wrote out a lengthy memorial in which he announced he was "transferring his allegiance to His Catholic Majesty."

So Wilkinson sold his cargo at a great profit and next appeared in Kentucky full of trappings of his new wealth. He laid out the future capital city of Frankfort and had a mansion built on Wilkinson Street. He also spoke highly of the Spanish as he resumed his quest for leadership.

Part of his plan required discrediting other leaders, and none of these was any bigger than George Rogers Clark, the Hero of the West. Clark was lionized for his capture of Vincennes in 1778, and action-loving Kentuckians looked to him for leadership. However, Clark's impetuousness in advocating attacking the Spanish played into Wilkinson's hands. By undermining Clark he could appease his new Spanish paymasters as well as hurt his chief rival for influence.

It was a Wilkinsonian trait to accuse others of vices with which he was passionately well-acquainted. Even before he went to New Orleans, he had started a whispering campaign that Clark had become a drunkard, charges that actually had some truth in later years. A series of anonymous letters were sent to various prominent citizens, all written in a style remarkably like Wilkinson's, that claimed Clark was unfit for command.

Wilkinson was able to discredit Clark, but he could not deliver a Republic of Kentucky into a Spanish Alliance. Ratification of the Constitution solidified the United States and led to Kentucky's admission as a state in 1792. By that time, Wilkinson was involved in the other major threat to frontier expansion: the Indian wars.

Indian raiding parties had always attacked Kentucky by coming down the Miami River. White raiders followed the same path and had noticed the fertile lands of what is now western Ohio. With Kentucky becoming more crowded, a spill over was natural and settlers were soon crossing the Ohio. The Indians had used the Ohio River as a settlement boundary since the time of Pontiac and now saw this threatened. Although some tribes were willing to concede land up to the Cuyahoga and Muskingum Rivers, most were willing to fight for the Ohio as a boundary.

The difference between Ohio and Kentucky was that by the time Ohio was settled there was a federal government -- although it remained to be seen whether it worked or not. Settlements north and west of the Ohio were made under the auspices of the Washington administration and therefore entitled to federal protection. But the entire United States Army consisted of a few hundred men under the command of General Josiah Harmar, stationed at the fort named for him across the Muskingum from Marietta.

In 1790, Harmar was ordered to take his force to Fort Washington near Cincinnati and mount a punitive campaign against Indian raiders. To do this he needed help from the militia, since he only had around 300 professional solders in his standing army. The militia were amateur citizen/soldiers who were called out in times of crisis to serve out short term enlistments. They were the Jeffersonian ideal except that they were poorly equipped, untrained and undisciplined. Any frontier military party needed a combination of regular troops and militia to succeed, but the state and federal forces often did not get along.

Harmar was able to add another 1100 militia to his force and left Fort Washington on September 30. As they slowly proceeded north, the Indians abandoned their villages and the soldiers burned them and the surrounding crops. This policy embittered the Indians while strengthening their resolve, and it frustrated the soldiers who had joined up to fight, not march. The army got as far as the present-day Fort Wayne, Indiana, before seeing any Indians. Colonel John Hardin, the veteran of Saratoga who had also come west after the war, was permitted to lead a smaller force in pursuit of combat.

On both October 21 and 22, the party was ambushed by a large Indian force led by Little Turtle and had to fight their way back with heavy losses. After this, Harmar returned to Fort Washington and pronounced the campaign a success because of crop destruction.

But the Indians felt they were the victors and resolved to press their advantage and step up frontier raiding. In January of 1791, they wiped out the settlement at Big Bottom, the most isolated of the Muskingum River outposts. They attacked Dunlop's Station near Cincinnati and burned a captive to death within earshot of the garrison. New settlement came to a standstill, and those already here were not safe once they were out of sight of their forts. Stronger military action was needed and this time Governor St. Clair was to lead the campaign personally.

St. Clair spent much of the year gathering men and supplies at Fort Washington, but many militia leaders and men did not turn out. The Kentucky militia were busy conducting their own slash and burn raids of Indian villages, and Wilkinson led one of these. St. Clair finally started north in October, and built advance posts at Fort Hamilton and Fort Jefferson. The lumbering army was hampered by inadequate supplies and miserable weather, and many men deserted. This became such a problem that St. Clair had to send his top regiment back to stem desertions and therefore his best troops were unavailable for battle.

Indian campaigns were often launched in the fall, when the harvesting tribes were considered vulnerable. The Harmar and Point Pleasant battles were in October and the St. Clair and Tippecanoe campaigns were in November. Another curious trait of the whites was when they won they gave the battle a name, such as Point Pleasant, Fallen Timbers, Tippecanoe and Wounded Knee. But when they lost, the fight was referred to as the Defeat of the name of the commanding officer. This explains the names of Lochry's Defeat, Crawford's Defeat, Harmar's Defeat and Custer's Last Stand. An addition to this is St. Clair's Defeat, the worst loss ever suffered by an American army.

On the night of November 3, 1791, the exhausted Americans set up camp in present day Mercer County, near the Indiana line. The regulars and militia slept in different camps and proper sentries were not posted. The next morning, a combined force of Indians, under the leadership of the Miami chief Little Turtle, attacked the camp. The Americans tried to rally from this surprise but were overwhelmed and the battle turned into a rout. Out of 1400 troops, the Americans had 630 killed and 283 wounded, or 65 per cent of their total. Also lost were all their cannon and many of the best officers. The defeat produced panic both on the frontier and in the administration.

Just before this disaster, Wilkinson had been offered a colonel's commission in the U. S. Army. This was a political move as it was felt it might help bridge the gap between the regular troops and the militia. Wilkinson was a popular Kentucky militia leader who that summer had led a mounted militia raid on Indian villages. But he was also a former general from the Continental Army with a reputation as a disciplinarian. He still was interested in the military and his finances were shaky despite Spanish largess, so he was only too eager for a fresh start.

Up to this time, Wilkinson's Spanish activities, while certainly unethical, were not treasonous. As a Kentuckian, he was free to seek the best deal for himself and his neighbors. But now he was about to become a high-ranking official in the new federal army. If there was an honorable time to sever his ties with Spain, this was it. Instead, he wrote New Orleans and demanded a raise, saying that he could be far more valuable to them in his new capacity. He also insisted that future documents refrain from using his name, but refer to him as Number Thirteen. This lucky number worked well for Wilkinson, for despite numerous rumors, his deception remained undetected in his lifetime.

Wilkinson came to the Northwest Territory early in 1792. He was given command of Fort Washington and promoted to Brigadier General in March after St. Clair and Harmar had resigned their commissions. He was considered for the post of commander-in-chief, but the job was offered instead to Anthony Wayne, another Revolutionary War veteran. Wilkinson's rank was second only to

Wayne's, and he wasted little time in trying to undermine his commander.

Wayne stayed at Fort Pitt, where he was trying to turn the remnants of the army into a disciplined fighting force. Wilkinson used his status as the ranking officer on the frontier as an excuse to correspond directly with the Secretary of War, and he used this forum to freely criticize Wayne and advance his own causes. Meanwhile, Wayne also had to deal with critics who opposed the concept of a standing army and with a low-caliber of recruits. He moved his American Legion downstream to Logstown and began intensive training.

Another factor working against Wayne was the unusual degree of Indian unity that he faced. Success had encouraged the various tribes, and making things worse was the Detroit-based British support they received. Detroit had been, since Pontiac's Conspiracy, the one British bastion that could not be taken. Clark had managed to capture the commander, the hated hairbuyer Henry Hamilton, but had never been able to take the post. Under terms of the Treaty of Paris, the British were not even supposed to be in Detroit, so their supplying and encouraging the Indians from there was particularly upsetting.

The frustrated Americans even considered negotiation with the allied tribes. In 1792, they sent out emissaries with peace belts to open discussions. Two of these men were murdered, including John Hardin. Only the delegation with Rufus Putnam and

John Heckewelder was unmolested, but the treaty they concluded at Vincennes did not include any of the important chiefs. In the meantime, Indian raiders grew increasingly bold and even hovered near federal military forts. In June of 1792, sixteen soldiers were killed in an attack on a haying expedition near Fort Jefferson, and a few months later, 11 more were killed when a pack train was raided near Fort St. Clair.

Wilkinson had built this last fort near present-day Eaton in March of 1792 as a way station between Forts Hamilton and Jefferson. He boasted that Fort St. Clair was sturdier than any of the other Ohio forts, and that it had been constructed quicker and cheaper and with fewer men, although he offered no figures to support this. One thing that was not cheaper was the series of homes that Wilkinson had built at all his forts. From the time he built his first Spanish-funded mansion at Frankfort, Wilkinson had shown a fondness for comfortable quarters that stood out on the frontier. He immediately had a home built at Fort Washington, and in 1792 he ordered another one built at Fort Hamilton. This two story frame house featured wooden floors, glass windows, a cellar, partitioned interior, and a two story veranda. If that didn't stick out enough amidst the plain pioneer architecture, the next year at Fort Jefferson Wilkinson had a quarters built that included a dormer and cupola. It is true that Wilkinson had his wife and small children to house, but his pattern of establishing ostentatious abodes at every post went beyond basic needs.

Wilkinson was in charge of all of these Ohio forts and tried to get them separated as an independent command. He continued these efforts even after Wayne had moved his army to Cincinnati in April of 1793. During this period he also continued to dispense advice to Spain and received compensation for it. He was nearly exposed when his agent was robbed and murdered by Spanish guides while hauling $6,000 in Spanish silver upstream to him. The culprits were captured and taken to a court in Kentucky where the judge was a Wilkinson crony. He ruled that this was somehow a military matter and remanded the prisoners to the commander at Fort Washington. Thus, Wilkinson got custody of the men who could expose him and he hustled them back to the Spanish, since the loss of the money, and his agent, was less important than the loss of his cover.

But Wilkinson's efforts were mainly focused on undermining Wayne, and in doing so he polarized the officer corps. After Wayne arrived in Cincinnati, Wilkinson spent most of his time at Fort Hamilton, where he was a magnet for disaffected officers. Since Wayne was as brusque as Wilkinson was charming, it wasn't hard to cultivate a coterie of cohorts. Wayne at first seemed unaware of or indifferent to Wilkinson's scheming, but when he saw the sorry state of supplies that Wilkinson was responsible for, he began to suspect his second-in-command of sabotage.

In spring of 1794, Wilkinson advanced beyond writing letters of complaint to individuals. He published an anonymous assault in the *Centinel of the Northwest Territory*, the only newspaper north of the Ohio, that claimed Wayne had lost the confidence of Congress and had refused peace overtures from the Indians. Federal officials wanted no part of a controversy, but Wilkinson appeared to be bringing matters to a head.

But after over two years of drilling, Wayne was finally ready to act. The previous winter, he had built posts at Fort Recovery on the site of St. Clair's defeat, and Fort Greeneville, where much of the army had spent the winter. Now he was going to use these advanced forts as a springboard to invade Indian country. The British had countered by building Fort Miami on the site of the Maumee Rapids -- a clear violation of the Treaty of Paris -- and garrisoning it with British troops.

Bolstered by this gesture, an Indian army went south in June to disrupt Wayne's supply lines. They ambushed a wagon train just outside Fort Recovery and then attacked the fort itself. This attack failed, in part because the Americans had found the cannon from St. Clair's army which the Indians had captured but hidden because they were not able to haul them away. Little Turtle's warriors had hoped to use these cannon against the fort, but instead found the guns were to be used against them.

This victory encouraged the American army, as did the arrival of 1500 mounted Kentucky militia in July. On the 28th of that month, Wayne's army marched from Fort Greenville. Instead of heading towards Indiana, they went straight for the heart of Indian country and the British fort. On August 8, he arrived at the confluence of the Auglaize and Maumee Rivers, the site of Pontiac's birth and many Indian gatherings. There, he built Fort Defiance and issued an ultimatum to the Indians, who asked for a delay. But Wayne continued to advance and the Indians were unable to surprise the well-trained troops, giving Wayne the reputation of a commander who never slept.

The tribes were still confident of British support and resolved to attack near Fort Miami, where a tornado had leveled some trees and formed a natural barrier. It was at these fallen timbers that the climactic battle of the Indian wars took place on August 20, 1794. The Indians attacked suddenly and scattered the militia, but there was no panic this time. The Americans regrouped and attacked and drove the Indians from the field with a bayonet charge. The whole battle lasted less than an hour, but it justified the years of preparation.

The defeated Indian warriors streamed back to Fort Miami, where they were shocked to find the gates barred. Wayne's army came on the scene and exposed the British as frauds. The Americans burned all Indian possessions around the fort while the British troops looked on helplessly, unwilling to risk total war. Wayne considered attacking the fort, but did not repeat Washington's mistake of 40 years earlier of starting an international war from an isolated frontier. It was a more effective victory to taunt the British and it exposed the shallowness of their professions of support for the Indians. All that came out of the fort was bombastic British bluster that rang hollow as the two commanders exchanged angry messages.

Wilkinson had been unable to prevent Wayne's victory, so now his only hope was to diminish it. Although he commanded a wing of the army and performed credibly during the battle, he wasted no time in attacking Wayne afterwards. In a correspondence to one of Kentucky's U. S. Senators, he complained of Wayne's lack of orders, his leaving of wounded on the field, and his conduct after the battle. He claimed he could have achieved the same result in a month with 1500 militia and complained of Wayne's "becoming every moment more secure and more inflated with his imaginary prowess and the importance of his puny victory."

As was the case with Harmar, it wasn't until later that it was seen who had really won, but subsequent events showed that there was nothing puny about Wayne's victory at Fallen Timbers. This became apparent the next summer when Wayne was able to gather all the leading chiefs at Greenville and basically dictate peace terms to them. The Treaty of Greenville opened up most of Ohio for safe settlement and opened the floodgates that led to statehood just eight years later. The Treaty was in fact, almost exactly in the middle of the territorial period and the watershed that made statehood possible. In the same year of 1795, England agreed to withdraw from the Northwest and give up Detroit, and Spain agreed to open the Mississippi for American commerce, thus removing all major barriers to settlement.

There were also repercussions from Fallen Timbers that affected all parties beyond the borders of Ohio. The British were effectively removed as a force on the American frontier, and except for the War of 1812, they were never again able to threaten American expansion. The Americans proved themselves capable of defending their frontiers, which made all future expansion possible. And the Indians had to face the fact that they had no allies they could play off each other anymore. They were forced to deal with the formidable Americans on their own. In fact, Fallen Timbers was the start of a 100-year period during which Indian lands were steadily lost to the encroaching United States.

Yet Wilkinson continued to try to wrest command away from Wayne, despite these obvious successes. He gave up trying to work through the Washington Administration, as it favored a trained standing army like the one Wayne had forged. Instead, Wilkinson aligned himself with the Jeffersonian faction in Congress, which favored militia as he did. He was unable to force the issue in Congress, however, and was in the process of asking for a formal investigation of Wayne's conduct that could tear the army apart when Wayne died suddenly in December of 1796. His death did what all of Wilkinson's machinations could not accomplish: it placed a Spanish spy in command of the U. S. Army.

For the next 16 years, Wilkinson held this position. During this time, he served under Presidents Washington, Adams, Jefferson and Madison, but he never fought a battle. He did, however, accept thousands of dollars from Spain, and since most activity took place in the Southwest, he continued to indulge in intrigue. He was constantly rumored to be in the pay of Spain, but no proof was ever found. After Spain ceded New Orleans to France, and France sold it to the U. S. in the Louisiana Purchase, it was Wilkinson who represented the country in the transfer ceremony. But he continued to advise Spain on how to hold onto Florida and Texas in the face of American expansion.

Though he never again served in Ohio, Wilkinson played a major role in the Burr Conspiracy that affected the Buckeye State. Wilkinson had known Aaron Burr since they invaded Canada together as teenagers in 1775, and in 1804 they renewed their friendship. At this time, Burr was serving as Jefferson's vice president, yet he was facing an uncertain political future. He had already announced he would not run for a second term, choosing instead to run for governor of New York.

But he had lost that election, partly due to the efforts of his rival, Alexander Hamilton.

On the night of May 23, 1804, Wilkinson came to Burr's New York City home and concocted a plan the exact details of which are unknown today, but it certainly involved Spain and the West. Not long afterwards, Burr got Jefferson to appoint Wilkinson as governor of Louisiana Territory, headquartered in St. Louis. Burr's brother-in-law was named Territorial Secretary and his stepson was given a judgeship in New Orleans. To further cement their alliance, Burr used his influence to get Wilkinson's sons accepted at Princeton, which was his alma mater as well as where his father had served as President.

Less than two months after this meeting, Burr killed Hamilton in a duel. Indicted for murder in New Jersey, he fled to Philadelphia, where he stayed with Charles Biddle, a cousin of Wilkinson's wife. With his eastern political career in ruins, Burr looked to the West as a new power base and possible source of empire. After his term as vice president was up in spring of 1805, Burr had a barge built in Pittsburgh and floated down the Ohio and Mississippi to New Orleans. He found that killing the Federalist Hamilton in a fair fight almost enhanced his reputation in the West, and he was feted wherever he went.

Burr traveled with John Smith, one of Ohio's first senators, and Jonathan Dayton, a former congressman and land speculator who gave his name to the city of Dayton. Their announced purpose was to explore the possibility of building a canal around the Falls of the Ohio at Louisville, but they were also sounding out frontier leaders with their schemes. Among the leaders Burr met with were future governor Thomas Worthington at Chillicothe and future Presidents William Henry Harrison and Andrew Jackson at their country estates. He also met in secret with Wilkinson at Fort Massac on the Ohio River, and rumors were already starting that their plan was to invade Mexico and wrest it from the Spanish.

One of the most important stops Burr made was at Blennerhassett Island on the Ohio, between Belpre and Parkersburg. Here he met wealthy Irish emigres Harmon and Margaret Blennerhassett, who had built a mansion and island Eden with money raised from the sale of their English estate. The Blennerhassetts claimed it was their support of the radical French Revolution that scandalized their neighbors and led to their move to America, but more scandalous was that the fact that Harmon had married his sister's daughter. The cultured couple was charmed by the erudite Burr and intrigued by the possibilities of his empire, and wound up financing a good portion of Burr's plans.

It is uncertain if Wilkinson was originally an enthusiastic participant or if he just strung Burr along before betraying him. Wilkinson continued to work for the Spanish while governor. He kept them informed of the progress of the Lewis and Clark expedition and even encouraged the Spanish to capture them. He also revealed his lack of respect for his patron Jefferson, referring to him contemptuously as "our fool". Wilkinson enjoyed controversial unlimited power as both governor and commander of the army, but his general's pay was only $225 per month. He needed Spanish gold to go with his American power, but in 1806, he almost had to choose.

In that year, a Spanish force crossed the Sabine River, which was the border between Mexico and the Louisiana Purchase. Wilkinson was ordered to take his army there and get the Spanish to leave American soil. Many on the frontier welcomed war with Spain and Burr saw this as an excuse to launch his plan under patriotic guise.

But Wilkinson had no desire to fight the hand that fed him. He stalled for as long as he could and he did have one excuse as his wife was dying at the time. But he finally arrived at the front months after being ordered there, and as American commander he had to figure out some way to avoid attacking the country he was spying for.

It was at this time he received the famous coded "cipher letter" from Burr. In it Burr reported his progress and promised to meet Wilkinson as he came down river. But Wilkinson now saw that by giving up Burr to both Spain and the U.S., he could save himself with both countries. He turned the cipher letter over to Jefferson and orders were given for Burr's arrest.

Meanwhile, Burr had been at Blennerhassett Island, where Blennerhassett had his Marietta business partner Dudley Woodbridge building boats for the expedition. Burr then blithely floated down the rivers, unaware that Wilkinson had betrayed him. He was arrested in Natchez and, after a failed escape attempt, was taken to Richmond, where he was to be tried for treason. Wilkinson used this crisis as an excuse to declare martial law in New Orleans, and he used martial law to arrest all those who possessed information that might put him in a position similar to Burr.

Those who think that scandal, bureaucratic stonewalling and circus-like trials are a modern phenomenon, need to hear about Burr's trial for treason. Held in 1807, it was the trial of its century. The judge was Chief Justice John Marshall, who was hearing the case in Virginia because the alleged attempt of treason had occurred on Blennerhassett Island, which was a part of Virginia. Marshall was a first cousin of Jefferson, but the two hated each other. Jefferson, who had already pronounced Burr guilty, stayed away but sent his protege, the recently returned hero, Meriwether Lewis, to monitor the trial and report back to him.

The foreman of the jury was Speaker of the House John Randolph, an enemy of Wilkinson's who called him "the most finished scoundrel that ever lived" and "the only man I ever saw that was from the bark to the very core a villain". Luther Martin led Burr's defense. A colorful former framer of the

Constitution, Martin kept a stone jug filled with whiskey beside him throughout the trial. Among the media hordes who descended on Richmond was Washington Irving and among the spectators was Andrew Jackson, who intentionally bumped Wilkinson from a public sidewalk and challenged him to a duel, which Wilkinson wisely declined.

Wilkinson was the star witness, the man called by one lawyer present "the alpha and omega" of this case. Yet he was characteristically late to show. Part of the reason was that he had to gather alibis. He got the Spanish governor of Florida to ship his archives to Havana so he could honestly write a letter saying he had no files in his possession that indicated that Wilkinson was receiving a Spanish pension. But Wilkinson damaged the case when under cross examination, he admitted to altering the cipher letter to avoid implicating himself. Prosecutor John Hay wrote to Jefferson of Wilkinson that "my confidence in him is shaken, if not destroyed." The jury agreed and nearly indicted Wilkinson, with the narrow vote going strictly along party lines.

Burr was acquitted after a brief deliberation, but the administration refiled charges in Ohio. This trial was never held, but the charges have not been withdrawn to this day. Though acquitted, Burr was ruined in this country, as were Blennerhasset, Smith and Dayton. Wilkinson, who was more guilty of treason than any of them, was returned to command of the army after a whitewashing court-martial of junior officers exonerated him.

He, at least, was not renewed as Governor of Louisiana Territory, as Lewis was named to replace him. Lewis was in over his head politically and died mysteriously of a gunshot wound in 1809 on his way to Washington to defend his expenses. His death was ruled a suicide, but at least one contemporary author believes Wilkinson may have had Lewis murdered to prevent him from exposing fraudulent practices during Wilkinson's tenure in office.

After returning to New Orleans, Wilkinson wasted no time before causing his next scandal. He moved the army out of town to a place called Terre aux Boeufs, which means Land of Cattle. But this swamp, which was leased at an exorbitant price from the man who would become Wilkinson's next father-in-law, was a pestilential hellhole. Drainage was poor and flies and mosquitoes were abundant. In addition, the provisions supplied, for which Wilkinson got a kickback, were rancid. It wasn't long before the troops began dropping like the flies that were all around them.

When the Secretary of War heard of this situation, he ordered Wilkinson to move the army immediately. Wilkinson ignored this communication and stalled when it was repeated, so that it took over four months of a delta summer before he could immediately move his army. In that time, nearly half his army had either died of disease, or deserted. At nearly the same spot six years later, Andrew Jackson would defeat a British army with only 13 men killed, yet Wilkinson was responsible for the deaths of nearly 800 men in peacetime while lining his own pocket.

Coupled with this scandal was a book published that had specific allegations of Wilkinson's involvement in the Spanish Conspiracy. Wilkinson was ordered to return to Washington and face another court martial. His luck held again as his defense counsel was Roger B. Taney, a brilliant young lawyer who would eventually succeed John Marshall as Chief Justice. Wilkinson was acquitted again and returned to command, despite the fact that he was spending more time in the courtroom than on the battlefield.

By now war with England was imminent and experienced soldiers were in demand. Wilkinson was promoted to Major General and sent to New York with orders to invade Canada. But his inability to work with others and to move quickly doomed his campaign. Except for his brief stint at Fallen Timbers, Wilkinson had not been on a battlefield since Saratoga, and he seemed to want to keep it that way. His troops couldn't help but notice how their General's health got worse as the enemy approached and how quickly he recovered after the battle. With their commander in his tent, the Americans were turned back in a poorly-fought battle and the opportunity to take Montreal was lost.

Wilkinson was replaced and ordered to face yet another court martial where he received yet another coat of whitewash. But by now the discredited general was no longer needed. After the war, he retired to New Orleans and wrote his memoirs: a self-serving three-volume set that drew little attention.

Wilkinson apparently had some schemes left, however, for he soon moved to Mexico City and began to lobby officials there for some unknown project. To provide a cover, he registered as an agent of the American Bible Society. It was at this time that he allegedly added opium to his list of bad habits. Nothing came of these plans and Wilkinson died in Mexico City in 1825. His death received little notice and it wasn't until many years later when the Spanish archives were opened that Americans could find out the full extent of the career of one of the biggest villains of the frontier.

CHAPTER VI
Pontiac, Joseph Brant, Tecumseh: A Trio of Natives

One of the many misconceptions about the Native Indians of North America is that they were united in their efforts to resist white encroachment. In actuality, the period before the arrival of Europeans was one of unrecorded but unremitting warfare among the various tribes whose own languages and cultures kept them apart. And when the whites with their advanced technology did arrive, the natives' first response was eagerness to use these advances against their tribal rivals.

In 1609, an Algonquin war party setting out against the rival Iroquois tribe invited French explorer Samuel de Champlain and his gun along. When Champlain and his companions shot and killed three chiefs with the new weapon, the bewildered Iroquois fled in terror and the Algonquin had a great victory. But the Iroquois were soon introduced to gunpowder by Dutch and English traders and for the next 150 years they waged fierce warfare against the Algonquin. As the various European powers struggled for control of land possessed by natives, the tribes aligned themselves with whatever group could best provide them with guns.

There were a few exceptions to this, as some farsighted chiefs came to realize that the fate of all tribes was threatened by ever-increasing numbers of whites. Three native leaders who first sought to unify the rival tribes were Pontiac, Joseph Brant and Tecumseh. All three, though ultimately unsuccessful, achieved an unprecedented degree of unity. And all three were born in what is now Ohio.

Pontiac was among the first to realize that Indian unity was necessary to stop westward expansion and his military alliance experienced great success in the spring of 1763, when nine of 11 most westerly posts were captured and burned. As was the case with all natives, there are no birth records, but it is believed that Pontiac was born at the confluence of the Auglaize and Maumee Rivers sometime between 1710 and 1720. He was the son of an Ottawa war chief and his mother was from another tribe, possibly the Chippewa or Ojibway. This probably helped give him a broader outlook.

Being the son of a chief guaranteed nothing, as the title was achieved by merit rather than heredity. The Ottawa had elder chiefs, or sachems, who specialized in diplomacy, but the title of war chief was attained by being successful in war. And a war chief continued to rise in stature as long as he had continued success, but found himself abandoned in failure. Pontiac proved himself a brave warrior and leader of raiding parties and rose to where he led the Ottawa contingent against the British at Braddock's Defeat in 1755.

But there is almost no record of him before 1763, when he suddenly became the most noted Indian on the continent. The first time he appears in any account is late in 1760, and even then he wasn't mentioned by name until a revised version appeared in 1765. This was when Pontiac encountered Major Robert Rogers at the mouth of the Cuyahoga in present-day Cleveland. Rogers and a small force were on their way west, having been sent the day after the British captured Montreal in September. With the capture of the last major French city in Canada, they decided to immediately occupy as many of the French posts along the Great Lakes as possible, as possession of these would strengthen England's bargaining position at the impending treaty talks.

The British generally held a low opinion of Americans as soldiers, but Rogers and his Rangers were an exception to this. Throughout the French and Indian War, this hand-picked force had specialized in lightning-quick raids on all terrain in all kinds of weather. Rogers was the logical choice to accomplish this mission before winter snows fell.

Moving along Lake Erie towards Detroit, Rogers was halted by a group of Ottawa braves at the mouth of the Cuyahoga on November 7. The braves were led by Pontiac, who demanded to know what they were doing on Indian land. Rogers explained his mission and Pontiac retired to consider his response. After a tense night, Pontiac in the morning announced that the whites could proceed as long as they treated him with respect and deference. Rogers then proceeded and took possession of Detroit on November 29.

This British victory meant tremendous upheaval for the Great Lakes Tribes, almost all of which had supported the French. They had traded with the French for years and had always been welcome at the various posts. The British trade goods were superior to the French, but the natives found themselves treated with disdain now that the former trading posts were British forts. The Midwestern tribes could not believe the French were giving up because of a few setbacks, and they were upset to be at the mercy of the British monopoly for the trade goods they had become dependent upon.

This feeling was exacerbated by the preachings of an Indian mystic called the Delaware Prophet. Natives placed great credence in dreams and the Prophet announced it was told to him in his dreams that the Indians could return to their glory if they would renounce all white contact and trade goods. It was completely unrealistic for the various tribes to give up the blankets, cooking pots and rifles that had transformed their lives, but the Delaware Prophet's message helped stir up anti-British feelings as they took control of the region.

There is some debate about how premeditated and organized the ensuing rebellion was and how much of a role Pontiac played in orchestrating it. But in the spring of 1763, an unprecedented coalition of 18 separate tribes disrupted the transition to British rule by attacking every western post in a campaign that came to be known as Pontiac's Conspiracy.

The fact that it was called a conspiracy shows that history is written by the winners, as a confederation of tribes designed to protect their land and way of life hardly deserves such a name.

In the spring of 1763, Pontiac was living in an Ottawa village on the Detroit River. Nearby were Huron and Potawatomie villages. It was advantageous to be located near the largest post in the Midwest, especially in spring when trading commenced. Major Henry Gladwin, the commander at Detroit, had 128 troops under his command, and the other scattered former French posts had much smaller contingents of regular British troops. In fact, of the nine other forts west of Fort Pitt, only Michilimackinac was garrisoned by more than 30 men and commanded by an officer with a higher rank than lieutenant.

On April 27, Pontiac addressed assembled tribes at the River Ecorces just below Detroit and outlined his plan. He was middle-aged and of medium height, but muscular and with dark complexion and long black hair. He excelled in strong oratory, which was important to the non-literate gathering. It was decided that the warriors would accompany Pontiac into the fort with filed-off rifles under their blankets. Pontiac would give a speech and during it would give a signal for the Indians to pull out their weapons and begin the attack.

However, the scheme was betrayed, either by an Indian maiden or someone who'd noticed a recent Indian demand for metal files. Pontiac found the garrison armed and ready, so he gave his speech, without giving the signal and left. He tried again later to gain admittance but Gladwin would only let small groups of Indians into the fort. So on May 9, Pontiac was forced to abandon trickery and begin the assault by attacking the village outside the fort's gates.

The siege of Detroit lasted six months, which was highly unusual for Indian warfare. While Indian warriors were known for individual daring and bravery, they normally did not have the discipline and patience needed for siege warfare. Indian braves preferred to strike suddenly when the odds were good and they could terrorize their foe and gain some plunder. It is a tribute to Pontiac's leadership and organizational skills that he could hold together such a diverse coalition for so long.

To supply his army, Pontiac was dependent upon the French Canadian farmers who lived nearby. To pay for goods commandeered from these neutrals, Pontiac issued promissory notes for payment on birch bark that bore the drawing of an otter, which was the sign of Pontiac's clan. It is said that all of these debts came to be honored.

But organizational skills were useless without military victories to feed the momentum. The Indians got one victory when they were able to surprise and capture most of a supply convoy on Lake Erie. The garrison was thoroughly demoralized when they discovered that their spring supplies had been captured, and since the plunder included whiskey, many of the captured were butchered in the drunken spree that followed. But more victories were needed to sustain the siege and the conflict gradually spread.

Fort Sandusky on Sandusky Bay was the closest fort to Detroit and the only post in what is now Ohio. Held by Ensign Christopher Pauli and only 14 men, this fort was taken on May 16, when the unsuspecting troops let Indians in for a parley. The entire garrison was massacred except for Pauli, who was taken to Detroit to be burned at the stake.

However, he was saved by being adopted by a widowed squaw, and he later was able to escape to the fort.

In the next month, forts from Erie, Pennsylvania, to Green Bay, Wisconsin, fell to the Indians, and soon only Forts Pitt and Detroit were still standing. Fort Miamis (Fort Wayne, Indiana) was captured when the commander was betrayed by his Indian mistress. At Fort Michilimackinac, Indians staged a competitive game of lacrosse just outside the fort's gates that the garrison watched. At a signal, the ball was thrown into the fort and the Indians rushed in and overwhelmed the garrison. It is not known what tricks were used to capture Fort Venango in Pennsylvania because no survivors were left to tell the tale. Isolated settlements and fur traders on the frontier were also targeted, with hundreds being killed or captured. Even Fort Pitt was besieged and all activity on the terror-stricken frontier came to a standstill. At Fort Pitt, desperate defenders stooped to giving Indians blankets that had been exposed to smallpox in a moderately successful early attempt at genocide by germ warfare.

The garrison at Detroit would have been completely cut off except for traditional British naval superiority. In taking the St. Lawrence River from France, the British were now able to sail all the way into the American heartland with only a brief portage around Niagara Falls. To alleviate the siege, relief ships could sail from Fort Niagara and dock at the fort's landing at the Detroit River. Indians tried to attack these ships in narrow channels of the river, but were no match for alert crews armed with cannons.

Late in July, a relief ship with fresh troops was able to land. Their commander, Captain James Dalyell, had a typical low opinion of Indians as fighters, and he convinced Gladwin that he could fight his way out of the fort. On the night of July 31, he tried this and was ambushed by Pontiac's men at Bloody Bridge not far from the fort. With Major Rogers' men covering the retreat, these men fought their way back to the fort with heavy losses.

Despite this Indian victory, morale sagged as the siege dragged along. Differences among tribes and disputes over Indian protocols concerning captives and plunder led to some breaks in Indian unity. Some chiefs let Gladwin know they had been forced into Pontiac's coalition and expressed a desire to make a separate peace. Pontiac's force was weakened by these defections, but he held out for the hope of French aid. This appeared to be an increasingly risky gamble, for if the French could not come through, the British would be the sole source of much needed gunpowder.

This is exactly what happened. As the first snows were falling in Detroit, Pontiac learned he could expect no help from the French. The commander at Fort Chartes, on the Mississippi, the last remaining French post in America, informed Pontiac of the terms of the just-ratified Treaty of Paris, in which France forfeited all American claims.

The next year the British sent out two expeditions to subdue the Indians. One force, under Colonel John Bradstreet, sailed the Great Lakes to relieve Detroit and blithely concluded peace treaties that it was not authorized to do. The other force was led by Colonel Henry Bouquet, the man who had lifted the siege of Fort Pitt by defeating the Indians at Bushy Run the previous August. In October of 1764, Bouquet led an unusually large force of 1200 from Fort Pitt to the Tuscarawas, then downstream to Coshocton, where he demanded the return of over 200 captives taken by Indians.

But Pontiac himself had still not sued for peace. Though he wintered on the Maumee, he began spending more time in the Illinois area, rallying tribes that were closer to French influence. British agent George Croghan found Pontiac in 1765 and persuaded him to come to Detroit in peace. The next year Pontiac represented the western tribes at a peace conference at Fort Ontario in Oswego, New York.

Here the British Supervisor of Indian Affairs, Sir William Johnson, treated Pontiac with great deference, which aroused jealousy among other chiefs. Once Pontiac agreed to peace terms, he honored them, which left him isolated from the war faction of his own people. Even though Pontiac had necessarily been co-opted by the peace conference, he was still viewed as a threat by both whites and Indians. On April 20, 1769, Pontiac was killed by a band of Peoria Indians while walking unarmed at a trading post in Cahokia, Illinois.

****

Any Indian leader who tried to unify the various tribes was bound to incur the wrath of white settlers, but few Indians were hated more by the Americans than Joseph Brant. As a leader of fierce raiding parties during the American Revolution, Brant was so adept at spreading terror that virtually every attack was blamed on him. His reputation as a perpetrator of atrocities was enhanced when he later killed his own son in a drunken brawl.

It was ironic that this alleged savage preached Indian unity by adapting white ways. Brant wrote and spoke English well, having been educated by the founder of Dartmouth College. He also converted to Christianity and helped to translate portions of the Bible into his Mohawk tongue. As a diplomat, he traveled to London to meet King George III and to Philadelphia to meet President Washington, and his travels included side trips to New York and Paris for shopping and research.

Although he often traveled in white circles, Brant did so only to secure the best deal for his people, and he never betrayed the circumstances of his humble Indian birth or his warrior heritage. As he said on a visit to London, "I like the harpsichord well and the organ still better; but I like the drum and trumpet best of all, for they make my heart beat quickly."

Brant's tribal affiliation enabled him to rise to prominence, but also helped sabotage his chances of achieving Indian unification. As a member of the Mohawk tribe, Brant belonged to the powerful Six Nations confederation. Also called the Iroquois League, this coalition of six tribes was the dominant force among eastern woodland Indians. The Six Nations were generally clustered around upstate New York, but so great was their strength that western tribes felt they occupied their present lands only because the Iroquois let them. The Six Nations lived in wooden houses, called longhouses, had picketed villages, practiced agriculture, and mixed easily with whites. Their method of strength through unification was noted by Benjamin Franklin, who used them as an example when urging the colonies to unite.

Brant's Mohawk tribe was the easternmost of the Six Nations, with the westernmost tribe being the Senecas, who ranged as far as eastern Ohio. There was much movement among tribes, particularly for hunting trips, and this is apparently how Joseph Brant came to be born in what is now Ohio. In late winter of 1743, his parents were living along a stream that was probably the Cuyahoga when they had a baby that they named Thayendanegea, which translates as "two sticks of wood bound together".

The family moved back to New York shortly afterward, where the father died. Brant's mother remarried but her new husband died, probably during an epidemic, and left the mother widowed again. Brant's humble birth made it unlikely he could rise to prominence among his people, but his fortunes were about to improve because of marital connections, traditionally a white method of advancement.

Around 1753, his mother married a Mohawk named Brant, who adopted her children and gave them his name. He was a sachem, or civil chief, of the local village, so the family's social status rose. And being a local chief meant hosting visiting dignitaries such as Sir William Johnson, who, as Superintendent of Indian Affairs for Great Britain, was possibly the most influential man in North America.

Johnson was a poor Irish immigrant who came to the Albany area in 1738 and began trading with Indians. He earned the respect of the tribes and was willing to live with them and like them, and he eventually became the diplomat most trusted by all tribes, although his loyalty was always to England first and the Iroquois second. Johnson had a common law wife but also several Indian mistresses, among them Joseph Brant's older sister, Molly. When his wife died in 1759, Johnson had Molly installed as the hostess of his estate, and she had several children by him.

Johnson's ability to work with Indians was invaluable in the French and Indian War, in which he became a Major General. His battlefield successes earned him a knighthood and extensive land grants, which were well-deserved as his efforts were a major reason for an Iroquois alliance and eventual British Victory. Johnson allowed his new brother-in-law to participate in his campaigns, so starting at age 15, young Joseph got his initiation as a warrior at places like Forts Ticonderoga and Niagara, and Montreal.

After these victories guaranteed British victory, Johnson took even more interest in young Joseph. In 1761, Johnson sent his protege to Lebanon, Connecticut, to be educated by the Reverend Eleazor Wheelock. Just eight years later, Wheelock would move to New Hampshire and found Dartmouth College, but his original goal was to educate Indians as potential missionaries. Joseph converted to Christianity, although his commitment was not unwavering. But he did take to education and soon learned to speak and write English quite well.

When Brant returned home, he was given useful work as a translator and interpreter, although he did not become Johnson's personal secretary, as some have claimed. But he was involved in most major Indian conferences, many of which were held at Johnson Hall, the suitably baronial mansion that Sir William had built. During this time, Joseph married and had children and was also involved in translating the Book of Mark into Mohawk.

Sir William Johnson was such a master of timing that he even died at the right time: right before his empire was about to crumble. In 1774, at Johnson Hall, he was giving an impassioned speech that helped successfully to keep the Iroquois out of Dunmore's War, thereby isolating the Shawnee and guaranteeing their defeat, when he collapsed and died. His son and son-in-law took over for him, but sentiment among the settlers of the Mohawk River Valley was generally for American independence, and those loyal to the crown were threatened.

The realignments that accompanied the American Revolution also ruptured the Six Nations, as the Oneida tribe sided with the Americans while the other tribes remained pro-British. Tensions between factions grew and in May 1775, the Johnsons evacuated Johnson Hall and fled to Canada, accompanied by Brant and loyal Indians.

There was considerable confusion about what to do next and a delegation was picked to sail to London for guidance. Because of his Johnson connection and ability with English, Brant was a logical choice to represent Indian interests. He spent six months in London, where he received promises of support for his people, a captain's commission in the British Army and an audience with King George the Third.

He also took full advantage of his novelty status and enjoyed a social whirl of masked balls and circuses, and he became a Mason. He had his portrait painted by Gilbert Stuart, an American studying in London, and was profiled by James Boswell for *London Magazine*. When Boswell asked what he liked about England, Brant replied that he "chiefly admired the ladies and horses".

He returned to New York in the summer of 1776 and rejoined his tribesmen near Fort Niagara, which would be their main base of operation for the rest of the war. The following year, Brant commanded the Indians who accompanied a British/Loyalist force that invaded the Mohawk Valley. The plan was to meet up with Burgoyne's army near Albany, but a stubborn American force at Fort Stanwix stood in their way. Upon hearing the reinforcements were heading to the fort, Brant's Indians set up on ambush.

On August 6, 1777, all of these former neighbors met at the bloody Battle of Oriskany. After a fierce day of fighting, Brant's warriors held the field but they and the British were considerably weakened. Two weeks later, the Americans were able to lift the siege of Fort Stanwix and the British retreated to Niagara, which made the American victory at Saratoga possible.

This was the last formal battle in the Mohawk Valley, but the warfare had just begun. Brant and Loyalist Walter Butler led brutal raiding parties that terrorized American settlements and earned Brant a reputation as a savage monster. He was blamed for massacres in the Wyoming and Cherry Valleys in 1778, even though he was not even at the first and tried to prevent indiscriminate slaughter at the second. But he was so effective as a raider that white Loyalists volunteered to come along and serve under him.

In 1779, the Americans sent to two large retaliatory expeditions that virtually destroyed all hostile Iroquois villages in the area. By the end of 1780, upstate New York was a desolate wasteland with few habitable villages left to raid.

The next spring, Captain Brant was sent west to Detroit to aid the western Indians. Even though he spent a full year in Ohio, New York settlers whose homes burned during this time still swore he had personally led raids against them. Brant's main contribution in the West was the capture of a 100-man relief party that was trying to catch up to George Rogers Clark's troops on the Ohio. After this victory, Brant met up with Simon Girty and the two most hated men on the frontier got drunk together. When Girty accused Brant of lying about his warrior prowess, Brant struck Girty with his sword and nearly killed him.

By the time of Cornwallis' surrender at Yorktown, the British realized the war would end soon, but no one knew what the terms would be. The Indians were shocked when the ratified Treaty of Paris said not one word about their fate. The Americans felt the Indians had forfeited their lands by aligning with the defeated British, and they claimed the land as a reward for their own troops.

Now more than ever, Brant saw the need for Indian unity. At an Indian conference at Sandusky in September of 1783, he said, "We, the Six Nations with this belt bind your hearts and minds with ours, that there may never be hereafter a separation between us, but there be Peace or War, it shall never disunite us, for our interests are alike, nor should anything ever be done but by the voice of the whole as we make but one of us."

The Americans continued to cross the Appalachians and pressed their claims with a number of coercive treaties signed by scattered non-influential Indians. In 1785, Brant resolved to return to London to try and enlist British aid. Though he was able to get a pension for himself and help with Canadian land grants for his displaced Iroquois, Brant got only vague assurances for Indians in general. He did, however, enjoy some shopping, and more of the London social whirl. He also arranged to have a Mohawk prayer book published, and took a side trip to Paris to see if the French archives had any additional information on the ancient Indian mounds that had intrigued him on his trips to Ohio.

The failure to get specific guarantees of aid from England made Brant skeptical of the English. He came to feel that the Indians were on their own and began to work harder for unity. But he also recognized the potential military might of the U.S. and advocated realistic concessions. The Ohio River had been the border between white and Indian for twenty years, but Brant felt it sensible to compromise and allow settlement to advance west to the Tuscarawas and Cuyahoga Rivers. Other tribes disagreed, including some that already granted concessions far beyond this. Proximity to the whites was a major factor in the diversity of Indian views, with the easternmost tribes obviously feeling the most threatened.

While the various tribes were in disarray, the Americans were already uniting to carve potential new states out of Indian territory. In 1787, the U.S. established the Territory Northwest of the River Ohio and named Arthur St. Clair as Governor. The following year, St. Clair took up office at the new settlement of Marietta and called for a conference to confirm previous treaty concessions. This was supposed to be held at the Falls of the Muskingum at present-day Duncan Falls. However, when a band of Indians raided the stores of presents and liquor that were an essential part of any treaty, St. Clair insisted the meeting be moved to Fort Harmar, across the Muskingum from Marietta.

The Indians were reluctant to meet under the guns of an American fort, and in the absence of a unified front, Brant recommended boycotting the proceedings. A few lesser Indians attended and accepted St. Clair's terms and presents, but the absence of tribes such as the Shawnee and Miami rendered the Treaty of Fort Harmar of questionable use.

Brant did consider attending and got as far as present-day Zanesville, where he wrote St. Clair in vain to request concessions. He sent his son Isaac to Marietta with the letter and out of this arose the story that Isaac fell in love with St. Clair's daughter, Louisa. Apparently Louisa was as headstrong as she was beautiful and in her determination to meet the infamous Joseph Brant she rode up the Muskingum on impulse, and Isaac could not resist her charms.

The story makes for a great legend, but there is no evidence to support it, and most likely the Brants returned home without meeting the governor's daughter.

As Indian raids by the more hostile tribes increased, the Americans decided to send their army out to punish them. Their objectives in the fall of 1790 were the Miami and Shawnee villages in the Maumee River area between present-day Fort Wayne and Toledo. This area was part of a water route from Quebec to New Orleans that only required two short portages and it was near the British port at Detroit. The Americans under General Josiah Harmar burned many unoccupied villages, but fled in battle when a portion under John Hardin was attacked. The Indians were so encouraged by their battle success that they increased their raids over the winter rather than wait until spring.

The next year a council was held at the Maumee Rapids. The pro-war faction was encouraged by their success and frowned on Brant's conceding of white settlement to the Tuscarawas and Cuyahoga. They resented an eastern outsider from the hated Iroquois advocating concessions after a victory. There was also some personal jealousy of the famed leader, for as Brant's former colleague Walter Butler had observed, Brant "has been more taken notice of than has been good for his own interest with his own people."

With Brant advocating waiting for more concrete assurances from the British, the warlike tribes allied under the leadership of the Miami chief, Little Turtle, an able war chief who had led them against Harmar. On November 4th, 1791, the Indians surprised a larger army under the command of St. Clair and nearly wiped it out.

St. Clair's defeat raised Indian confidence to a new high and led the Americans to seek conciliation. One way of doing this was an attempt to co-opt Brant and get him to use his influence on their behalf. During the Revolution, Washington had considered a plan to kidnap Brant, but now he hoped to win him over. Governor Clinton of New York wrote President Washington that Brant "is a man of very considerable information, influence, and enterprize and ... his friendship is worthy of cultivation at some expense."

As a diplomat, Brant was willing to talk to anyone who might be able to help his people, and in June, 1792, he came to Philadelphia to meet with Washington and Secretary of War Henry Knox. The Americans basically offered him bribes to use his influence with the western tribes. But Brant was not interested in bribes and his influence was less than the Americans realized.

At this same time, an Indian council was being held at the Glaize: the high ground where the Auglaize and Maumee Rivers meet in what is now Defiance. This was one of the largest Indian councils ever held, with Simon Girty the only white allowed to attend. The war faction was clearly in ascendancy now, and Brant could not have done anything to stem this tide, especially since some resented his lack of participation in recent battles and his consorting with the Americans. As it was, Brant was delayed by illness and did not arrive at the Glaize until just after the conference had adjourned.

Plans were made for a conference the next year at the Maumee Rapids that the U.S. commissioners were invited to. Brant was enthusiastic about this, saying, "This is the best time to obtain a good Peace, and if lost, may not be easily re-gained." But when he arrived before the conference in the spring of 1793, he found the western tribes trying to isolate him and the Iroquois. The war faction plan was to demand terms so harsh that war would be inevitable.

But the Indians found they still needed Brant's presence as well as his diplomatic skills and facility with English, and they had to give him a prominent role. He was able to keep the talks going for a while, but Indian insistence on no compromise on the Ohio River boundary doomed his efforts and the talks collapsed in late summer.

Over the winter, the British talked more openly of support and early in 1794, they backed this up by building Fort Miami at the Maumee Rapids. This was a clear violation of the Treaty of Paris and convinced the Indians that the British were willing to go to war alongside them. Unfortunately, they found out otherwise after the ensuing Battle of Fallen Timbers, when British support evaporated in the face of Anthony Wayne's well-trained army. The Indian war coalition led by Little Turtle was one of the most successful, but they would have been better off listening to Brant. In 1795, Wayne was able to dictate peace terms at the Treaty of Greenville and the Indian opportunity to obtain a good peace was never regained.

After this defeat, Brant concentrated his efforts on improving the lot of his own Mohawks in the Grand River area of Ontario. He endured tragedy in 1795 when his dissolute son Isaac attacked him in a drunken rage and Brant killed him in self-defense. To his many detractors, killing his first-born son only cemented Brant's reputation as a monster.

But Brant's final years were peaceful ones. He was able to secure permanent ownership of land for his people and tried to get them to adopt white ways, particularly in agriculture. The Grand River town of Brantford, Ontario is named for him.

In his last years, Brant built a mansion that was vaguely reminiscent of Johnson Hall. He died in this home in 1807. It is appropriate that he died peacefully at home, for despite being known as a warrior, he was actually a skilled diplomat.

****

Of all natives who attempted to unite the various tribes, Tecumseh was the most impressive, and he came the closest to success. All contemporary descriptions of him mention his striking

Gliding to a Better Place

looks and imposing presence, his speaking abilities, as well as his skill as a warrior. He toiled relentlessly for Indian unity despite all the barriers that stood in the way of his mission, but his movement died with him.

Tecumseh's tribal and family background shaped his movement. The Shawnee tribe was highly nomadic as evidenced by the fact that Tecumseh's mother was born in Georgia, and died in Missouri, though Tecumseh spent his life in Ohio and Indiana. The Shawnee therefore came into contact with many other tribes, and as fierce warriors they were not afraid to take on the mighty Cherokee or Iroquois.

Like all tribes going back as far as the Old Testament, the Shawnee had their own language, culture, and religion that placed themselves at the center of the world. If the tribe suffered reverses, their misfortune would be interpreted as their Creator's way of expressing dissatisfaction of their straying from the proper spiritual path.

Tecumseh's parents met while they were in Georgia, but they came north in the late 1750s in search of good hunting grounds. The Shawnee who migrated at this time used Kentucky as hunting grounds, but built their villages north of the Ohio, along the Scioto, Miami, and Little Miami Rivers.

Tecumseh was born in what is now Ohio around March of 1768, but the exact location in unknown. Most believe the birthplace to be along the Little Miami near Xenia, but it also could have been at the village of Chillicothe on the Scioto. And in 1807, Tecumseh was traveling with two future governors of Ohio and when they passed a spot on the Mad River just west of Springfield that Tecumseh said was his birthplace. However, this spot was a later Shawnee town where Tecumseh lived as a boy, but was not founded until after his birth. The lack of written records make it difficult to know the exact location or even the significance of Tecumseh's name, but the most likely translation is "Celestial Panther who waits for his Prey". It is known, however, that the "s" in Tecumseh's name was pronounced with a " th" sound.

Tecumseh was the fourth child in the family, with the oldest son, Cheesekaw, having been born in 1760. The family was rounded out in 1775 by the highly unusual birth of triplets. But by then, the family had been torn asunder by the death of Tecumseh's father in the first of many tumultuous conflicts that engulfed the tribe during Tecumseh's brief lifetime.

At the Treaty of Fort Stanwix in 1768, the year of Tecumseh's birth, the Iroquois had given permission for whites to settle in Kentucky, even though the land was used by the Shawnee. This treaty isolated the Shawnee and they found few allies when they tried to stop the new white settlement. In 1774 they fought the Virginians and Tecumseh's father was killed at the Battle of Point Pleasant. Cheesekaw, who also fought at that battle, became a surrogate father to his six-year-old brother.

Not long after the Shawnee were forced to accept harsh peace terms, the American Revolution began, which plunged the tribe into more war. The Shawnee were originally neutral but when their principal chief Cornstalk was murdered while on a peaceful mission to the Point Pleasant site, they were driven to the British side. The Shawnee were responsible for most of the raids on the Kentucky settlements, which led to retaliatory raids of their own villages.

In May of 1779, a party of Kentuckians attacked the new village of Chillicothe, not far from Tecumseh's village, and killed their chief, Blackfish. The next year, a party of Indians and British attacked Kentucky and returned with many prisoners, one of whom was adopted into Tecumseh's family. At twelve years old, Stephen Ruddell was the same age as Tecumseh. This form of adoption was often used by tribes to replenish their ranks. In fact, Daniel Boone had been captured and adopted by Blackfish and remained with him for six months in Ohio before escaping to warn Boonesborough of an attack.

In the next round of raids, George Rogers Clark led a force that August to Tecumseh's village near Springfield. Clark burned the village and forced the Shawnee farther north towards Piqua and Bellefontaine. Tecumseh was twelve at the time of this attack and probably not involved as a warrior. It is possible he got his first taste of battle two years later at Blue Licks in Kentucky, when Kentuckians fell into a trap set by Simon Girty despite being warned by Boone. However, it is not certain he fought here either.

After the Americans gained independence from Britain, they increased pressure on Indians for land concessions. The Shawnee at first boycotted all negotiations, but they were summoned to Fort Finney at the mouth of the Great Miami in January of 1786. Here they were told they must give up their claim to the Miami Valley or face war. Those present reluctantly agreed, bu t the rest of the outraged tribe refused to acknowledge the treaty and warfare broke out anew. In October of 1786 a Kentucky party attacked and destroyed a village near current West Liberty. The chief, Moluntha, was murdered in cold blood even though he had been one of the chiefs to sign and honor the treaty.

Tecumseh had attained warrior status by this time, although a legend exists that said he fled at his first taste of battle. The young warrior had grown up handsome and serious, and was an excellent hunter. Under the tutelage of his brother, he began to accompany raiding parties along the Ohio River. The outnumbered Indians were hesitant to risk losses in open battle, and preferred to lure individual white boating parties ashore and attack them. It was after some captives of one such venture had been burnt that Tecumseh first spoke out against torture.

While much that has been said of Indians as noble savages is true, it should not be forgotten that this was a people for whom public execution by torture was considered grand spectator sport.

Captives were part of structured Indian protocol and could be burned for revenge with religious implications. But while Tecumseh had no sympathy for the whites who killed his father, burnt his village, and murdered his chiefs, he had no tolerance for torture of a helpless foe.

Around 1788, Tecumseh left Ohio with a small band under the leadership of Cheesekaw. This group first went west to Missouri, where many Shawnee, including the boys' mother, had relocated. Later they went to Tennessee where they lived off the land and participated in many raids on white settlers. It was on one such raid that Cheesekaw was killed, and after avenging this death by taking several scalps, Tecumseh returned north at 1791.

Tecumseh returned to the Ohio country to join one of the most successful tribal coalitions. Led by the Miami Little Turtle and the Shawnee Blue Jacket, the allied Indians held the U.S. Army at bay for nearly five years. Tecumseh led a group of Shawnee scouts who harassed and spied on the troops, but he did not participate in the rout of St. Clair's Army in 1791.

He preferred to work with small speedy raiding parties that terrorized settlers. On two occasions, his group was followed and attacked by a band led by Simon Kenton. Near present day Williamsburg, in spring of 1792, and near current Bainbridge the following spring, his group was attacked. Both fierce fights were to a draw, with Tecumseh able to fight his way out and escape.

With the selection of Anthony Wayne as U.S. commander, the Indian luck ran out. Wayne drilled and trained his troops extensively, and even Little Turtle realized he could not take this army by surprise. In 1794, a large Indian army resolved to strike at Wayne's supply line. Tecumseh was in this force, which attacked a large convoy outside Fort Recovery on June 30th. They were able to overrun the wagon train but then impulsively attacked the fort, where they were beaten back with heavy losses. Having taken both plunder and scalps, many Indians left the coalition and they had fewer men to face the Americans as Wayne advanced.

Tecumseh was also at the Battle of Fallen Timbers, where the Americans were surprised but did not panic. And after the defeat, he saw that the British promises of help were empty. When Wayne summoned all tribes the next year to Greenville, Tecumseh did not go to the treaty negotiations. Here Wayne dictated terms that not only opened up most of Ohio to settlement, but allowed the U.S. to keep Forts Defiance and Wayne in the middle of Shawnee country.

For the next ten years, Tecumseh lived in different locations with several followers. For a while, his town was along Buck Creek in Clark County, and he also lived along the White River in Indiana. He moved freely in an area that whites were settling and got to know some of the newcomers on a friendly basis. Among the pioneers he knew was Albert Galloway, who arrived near Xenia in 1797.

From this connection, there arose a legend that Tecumseh fell in love with Galloway's daughter Rebecca. Tecumseh was alleged to have proposed, but broke it off because Rebecca insisted he live among whites. Most likely, this was a story told later by an older Rebecca Galloway to entertain her grandchildren, for there is no evidence of its truth and it does not fit with what is known of Tecumseh.

It is true, however, that the mother of Tecumseh's one son was half-white. Apparently Tecumseh married a half breed named Mamete, but this liaison was brief, and Tecumseh entered into it because he felt a chief was supposed to have a family. For the most part, he had little interest in romance, and was totally involved in providing for his people.

Tecumseh was seriously earnest in this quest, although he could be capable of humor. As a leader, he was sensible, generous, proud, and honest to a fault. He deplored any sign of elegance or effeminacy. Most importantly, he had a charismatic presence that could not be ignored, although at this point few whites were aware of him.

A contrast to Tecumseh was his brother Lalawethika, one of the triplets born in 1775. Whereas Tecumseh was known for his striking good looks, this brother was disfigured from a childhood accident that left one eye closed. Lalawethika was a medicine man, not a hunter or warrior. He was also an extravagantly boastful profligate wastrel who drank himself into a coma in 1805. But while he lay near death, he had a dream that transformed the lives of both brothers and led to a movement that affected the entire frontier.

Indians put great stock in dreams bearing spiritual messages. Prophetic interpreters of dreams played a role in political affairs from the time of Pontiac's Conspiracy to the Ghost Shirt movement that culminated at Wounded Knee in 1890. When Lalawethika woke from a coma so deep that burial preparations had begun, he announced his dream vision and launched a reform movement under his new name, the Prophet.

His message was similar to the Old Testament prophets: that the Creator had withheld blessings to His chosen people because they had strayed from the righteous path. To return to their glory years, they needed to shun the devil whites and the products with which they had corrupted Indian culture. Foremost among these was alcohol, but also proscribed were beef, pork, white fibers and extravagant dress, and white bread. Guns could only be used in warfare against whites--hunting had to be returned to the realm of bow and arrow. All property was to be communal and false prophets were to be condemned.

The startling religious pronouncements just happened to coincide with Tecumseh's political agenda and henceforth the brothers had sort of a symbiotic relationship. The movement got attention quickly and when skeptics challenged the Prophet, he gained credibility by claiming credit when a solar eclipse occurred. He was also not above a witch hunt to advance his cause. White-induced disease was decimating many tribes at this time and the superstitious were looking for a cause. When local Delaware accused some of their own of witchcraft, the Prophet came to confirm this diagnosis, and five tribesmen were burned. He later tried the same with the Wyandots, but their chief, Tarhe, put a stop to it.

The brothers' movement experienced success among many tribes, but caused a schism in their own tribe. In 1806, they moved their town to the site of abandoned Fort Greeneville, in accordance with a vision of the Prophet's. Wapakoneta was the main Shawnee town, under the leadership of Black Hoof, and a great rivalry resulted. Black Hoof was of the traditional leadership clan, yet he more progressively believed the Shawnee should be practicing agriculture and imitating the whites' ways if they wanted to survive. Tecumseh and the Prophet were young upstarts, yet held more reactionary positions. They felt that the Black Hoof faction was corrupt and dependent on the Americans, and considered agriculture to be work suited for women, not warriors.

In addition to causing consternation among Indians, the brothers' movement was noticed by whites. Among the first to become alarmed was William Wells, the U.S. Indian agent stationed at Fort Wayne. Wells, who was Tecumseh's age, had been captured in an Indian raid of Kentucky and raised by the Miamis. He married Little Turtle's daughter and fought against both Harmar and St. Clair. However, he then went back to the white side and gave Wayne valuable information and scouting help. Wells felt the new movement was a real threat to frontier security and Tecumseh in turn regarded Wells as the worst kind of corrupt white.

By 1807, a steady stream of Indian pilgrims visited Greeneville and local whites grew increasingly uneasy. In May, a settler was killed by Indians, and the Wapakoneta and Greenville Shawnee blamed each other. A conference to discuss the issue was held at Urbana and here Tecumseh met his old foe, Simon Kenton, and they reminisced about when they had tried to kill each other. At another meeting in Springfield, the personable Tecumseh stayed to successfully compete in sports and games with locals.

But fears still persisted as far away as the state capitol of Chillicothe, and in September, Governor Thomas Kirker sent a delegation to Greenville led by future governors Thomas Worthington and Duncan MacArthur. They returned to Chillicothe with Tecumseh and a few other chiefs, among them Blue Jacket, whose support had lent credibility to Tecumseh's movement.

Though Tecumseh spoke some English, he didn't want to be misunderstood or mocked, so he used a translator when speaking in public. At Chillicothe, his translator was his childhood companion, Stephen Ruddell, now a repatriated minister. On September 19th, Tecumseh gave an impressive yet reassuring address that won over the

Ohioans completely, renouncing all military intentions and stressing the religious nature of the movement. He furthered his good first impression on white society by remaining a few days at Adena, the new stone mansion built by Worthington just outside town.

But the propinquity to white society was going to remain a problem, and in 1808 the brothers relocated to Indiana territory and built Prophetstown near the junction of the Wabash and Tippecanoe Rivers. The new town prospered immediately and attracted members from the many Midwest tribes, but isolation from the whites was going to be impossible.

Governor William Henry Harrison of Indiana Territory was pursuing an aggressive land acquisition policy in order to attract enough settlers to obtain statehood. He instigated a series of treaty negotiations that further encroached on Indian lands. This pushed the Indians towards the British, with whom the Americans increasingly came to feel they would be going to war. At first the brothers attempted to be conciliatory, with the Prophet going to the Territorial capital at Vincennes to reassure Harrison while at the same time Tecumseh was negotiating with the British at Fort Malden, across the river from Detroit.

But in 1809, Harrison bribed and coerced several minor tribes into signing the Treaty of Fort Wayne, which gave up lands these tribes had little or no claim to. This enraged Tecumseh, who espoused the Dish with One Spoon concept, where all lands belonged to all tribes, and no one could sell land without the consent of all. He denounced the treaty as void, and declared surveyors of the lands would not be safe. He also stated that the Indians wanted only fair trade with the whites, not gifts like the annual distribution of salt that had become a charity that the Indians had become dependent upon.

Harrison replied that Tecumseh was welcome to go to Washington and appeal to the President if he wished, and invited him first to Vincennes to talk. Tecumseh went there in August of 1810, but mindful of what had happened to previous Shawnee chiefs under flags of truce, he came with a retinue of several hundred braves. The two leaders met at Grouseland, Harrison's brick mansion. Tecumseh refused to have negotiations inside, saying, "the earth is my mother, and on her bosom I will recline."

In an impassioned and dignified speech, Tecumseh recounted the history of white lies to his people and openly threatened the chiefs who had sold out at Fort Wayne. He again stated the Indians were to be treated as one people, and used the United States as an example of a federalized coalition. When Harrison refuted this notion, saying the tribes could not even agree on languages, and then defended his treaty, Tecumseh became enraged and called Harrison a liar. His braves reached for tomahawks, Harrison drew his sword, and violence would have ensued had not Territorial Secretary John Gibson, who spoke Shawnee, realized the seriousness and had a guard of soldiers

Gliding to a Better Place

brought up in time. The talks were suspended for the day and even though Tecumseh apologized the next day, little was accomplished.

Harrison didn't understand Tecumseh's position, and there was no way he was going to give up land he felt he had a title to. However, he was a perceptive man, and having met both brothers, he realized that it was Tecumseh who was "the Moses of the family." Tecumseh now began to prepare for war, and he went among all Midwestern tribes to rally support. Harrison had this to say about his foe and the efforts he made:

"The implicit obedience and respect which the followers of Tecumseh pay to him is really astonishing, and much more than any other circumstance bespeaks him one of those uncommon geniuses which spring up occasionally to produce revolutions and overturn the established order of things. If it were not for the vicinity of the United States, he would, perhaps, be the founder of an empire that would rival in glory that of Mexico or Peru. No difficulties deter him. His activity and industry supply the want of letters. For four years, he has been in constant motion. You see him today on the Wabash. In a short time you hear of him on the shores of Lake Erie or Michigan, or on the banks of the Mississippi, and wherever he goes, he makes an impression favorable to his purposes."

Indeed, the indefatigable Tecumseh was everywhere, meeting with everyone from the Creeks of Georgia to the Osage of Missouri and the Sioux of

Minnesota. But there were problems inherent with Indian unification that made his mission nearly impossible. For one thing, Indians tended to be jealous of a leader who got too powerful, yet without strong leadership they were ineffectual. And a more obvious problem was overcoming the language and geographical barriers and the traditions of hundreds of years of intertribal warfare.

The outnumbered Indians could not spare warriors, so they tended to be frontrunners who would support a cause only when victory seemed certain. And since the situation could change easily, Tecumseh constantly had to revisit allies to reaffirm alliances. And even among those who seemed favorable to him, he still had to restrain the young war hawks from attacking before the time was right. Another problem was chiefs who had become addicted to white goods and threatened Indian solidarity by selling out. Yet Tecumseh needed white help also, because his people might have had guns but lacked the ability to repair them or a steady gunpowder supply. This drove him to seek British support, even though he knew the British just wanted Indians to fight their battles for them.

Tecumseh felt war was inevitable. Still, he made one more visit to Vincennes in 1811. Immediately afterwards, he left for a tour of southern tribes that took several months. He had mixed results there, but legends persist about how he threatened to destroy a village that spurned his efforts shortly before the New Madrid earthquake wiped out every building there.

But Tecumseh made a serious mistake when he openly announced to Harrison that he would be leaving for a lengthy southern trip. Harrison had managed to convince federal officials that sufficient dangers existed to authorize aggressive action, and he realized that with Tecumseh gone, it was the perfect time to attack. He raised a 1000-man-army in the fall of 1811 and marched on to Prophetstown. They got within a few miles of the town on the night of November 6th.

It was the Indians who attacked first, however, since the Prophet had promised them a great victory. But the Americans withstood the assault at the Battle of Tippecanoe and destroyed the town as well as what was left of the Prophet's credibility. Tecumseh was furious when he returned, especially since he had told the Prophet to avoid battle with the whites. That winter, Indian raiding was the most active since 1791: Harrison's aggressiveness had not ended a war, but started one.

Tecumseh continued to rebuild and negotiate for peace, appearing at a conference in Fort Wayne in spring of 1812. But when war was declared between the United States and Britain, Tecumseh was already at Fort Malden, where he offered the services of himself and his warriors to British General Isaac Brock. The outnumbered Brock made good use of Tecumseh's men and through aggressive action convinced a superior force of Americans under

General William Hull to surrender Detroit without a fight on August 16. At almost the same time, Fort Dearborn (Chicago) was evacuated and many of the garrison massacred, including William Wells. Fort Wayne was able to withstand attack, but for the most part, the fall of 1812 was a promising time for Tecumseh.

Unfortunately, Brock was killed in battle in October near Niagara Falls, and Tecumseh's chance for ultimate success may have died with him. The two leaders had enjoyed a good rapport, with Tecumseh saying of Brock, "this is a man". Brock called Tecumseh "the Wellington of the Indians", adding that "a more sagacious or more gallant warrior does not, I believe, exist." Even more importantly, Brock showed every indication of honoring Tecumseh's request for an Indian buffer state between the U.S. and Canada. Brock's successor, Colonel Henry Procter, suffered greatly in comparison to Brock.

After such a disastrous beginning, the U.S. named Harrison as its western commander, and he advanced from Fort Wayne to the Maumee Rapids, where he built Fort Meigs and named it for Ohio's governor. In the spring, the British and Indians marched down from Detroit to attack, but the Americans did not come out to fight. With the American army holed up in the well-designed fort, artillery bombardment was all that could be done, and Tecumseh complained that he couldn't fight men when they acted like ground hogs.

But they got their chance to fight on May 5, when a large force of Kentucky militia rushed to reinforce the fort. A portion of them successfully attacked the British batteries, but then made the mistake of following a band of Indians into the woods. They were led into an ambush and nearly all were killed or captured. Indians began to butcher the captives, but an outraged Tecumseh rode up and stopped them, thereby saving hundreds of American lives.

Having had their battle and taken their fill of plunder and scalps, many Indians left the army, and the siege had to be abandoned a few days later. Procter retreated to Detroit, but in July Tecumseh convinced him to try another attack. But again the American groundhogs could not be lured out of their fort, and the British decided to attack the smaller Fort Stephenson nearby.

This fort was held by 170 men and one cannon under the command of 23-year-old Major George Croghan, a nephew of George Rogers Clark. Though outnumbered by more than 10 to 1, they refused to surrender and repulsed a bloody frontal assault on August 2nd.

This was a psychological victory for the United States, but a very tangible victory was soon won by Commodore Oliver Hazard Perry, who had been building a fleet of ships at Erie, Pennsylvania, to challenge the British fleet on Lake Erie. Control of the Lake was essential, since fleets could land behind the enemy with troops to disrupt supply lines. The

British were used to naval supremacy, but on September 10th, 1813, Perry's fleet met them at Put-In-Bay, and after a bloody four hour battle was able to report "We have met the enemy and they are ours."

Back at Fort Malden, Procter failed to tell his Indian allies about this key naval defeat and then tried to secretly organize a retreat. A furious Tecumseh publicly questioned Procter's manhood and demanded they make a stand, which Procter agreed to do somewhere along the Thames River. While retreating from Harrison's army, Procter was far in advance, while Tecumseh calmly remained until all of his braves and their families were safe and the Americans were in sight.

The British finally made a stand near Moraviantown on October 5th. But they were greatly outnumbered and fled after a brief battle, leaving their Indian allies to fend for themselves. Sometime during this part of the battle, Tecumseh was killed. The details of his death and the fate of his body remain controversial. But in 1836, Colonel Richard Mentor Johnson was elected vice-president campaigning on a slogan that claimed he had killed Tecumseh. Then in 1840, Tecumseh's main foe Harrison was elected president on the slogan, "Tippecanoe and Tyler, Too". Such was the strength of the legend surrounding Tecumseh that he was still a force in national politics over 20 years after his death. There was even a Messianic myth that said Tecumseh would return and all Indians would unite at his second coming.

At the end of the war of 1812, the British did propose an Indian buffer state, but the Americans were now in a stronger bargaining position and were able to sweep this plan off the table. The Indians never again came as close to a successful unification as they did under the leadership of Tecumseh.

CHAPTER VII
Abraham Whipple: The Last Voyage of the Commodore

On a late April day in 1801, a crowd gathered near the confluence of the Ohio and Muskingum Rivers in Marietta. The center of attention was a two-masted ship at the water's edge, approximately 2,000 miles from the nearest reachable ocean. The crowd had gathered to watch the launching of the brig *St. Clair* as it embarked on a journey from Ohio to the West Indies. This voyage, which would rescue the economy of the young settlement, was a bold venture to undertake. But just as bold was the man to whom the owners had entrusted care of the ship: Commodore Abraham Whipple.

For Whipple, this voyage would complete a trio of naval firsts. In 1775, Whipple became the first to fire on and capture a British ship in the American Revolution. In 1784, Whipple became the first to triumphantly sail the Stars and Stripes on the Thames to London after Britain had been forced to accept American independence. Now, at age 68, Whipple was about to become the first to sail a square-rigged vessel down the Ohio and Mississippi Rivers.

Commodore Whipple's last voyage capped the fascinating career of a true hero of the Revolution and spawned a brief boom in an inland shipbuilding industry in the years immediately before development of the steamboat. And while wood for ships and hemp for rope were readily available in the Mid-Ohio Valley, Whipple was the only man available who could have gotten the ship safely to sea.

Although he had been a landlocked farmer for the past seventeen years, Whipple was a natural seaman. Born in the port city of Providence, Rhode Island in 1733, Whipple was descended from an associate of Roger Williams, who founded the colony. Young Abraham took an early interest in the sea and also managed to teach himself navigation and bookkeeping.

Whipple's leadership abilities must have been obvious early on, because he was captain of his own ship while still in his 20's. During the French and Indian War, he was one of the most successful privateers in the American colonies. Under the privateering system, independent sailors with "letters of marque" from governmental authorities could commit legalized piracy on a designated foe. This subcontracting mercenary method was a risky but potentially lucrative endeavor, particularly with a daring captain, which Whipple certainly was. According to a notice in the *Boston Post Boy and Advertiser* from February 4, 1760, Whipple's ship *Gamecock* captured twenty three French ships, or prizes, in a seven month cruise.

There is an early illustration of Whipple's nautical daring from this period. While in the Caribbean, a storm forced Whipple to jettison most of his cannon to lighten his load. Shortly after this, a French ship appeared and Whipple was forced to flee, something he was loathe to do. However, he soon employed the old trick of using "Quakers" -- blackened logs that from a distance looked like cannon. He stopped short and showed fight and the French ship fell for the bluff. Not content merely to have escaped, Whipple gave chase like a dog after a truck until the fleeing French ship was out of sight.

In 1761, Whipple married Sarah Hopkins, who had Rhode Island connections as strong as his own. One of her uncles, Stephen Hopkins, later became Governor of the state and was also a member of Continental Congress who signed the Declaration of Independence. Another uncle, Esek Hopkins, was to

Gliding to a Better Place

become the first commander-in-chief of the American Navy. The couple took up residence in Providence and had two daughters and a son. After the French and Indian War, Whipple stayed at sea and became involved in the West Indies trade.

Oppressive British trade policies eventually got Whipple into conflict again. Crown policy dictated that all imports had to be in British ships with predominantly British crews. Outraged New Englanders turned to smuggling in great numbers. The British patrolled the coast with ships like the schooner *Gaspee*, which had orders to stop and search colonial vessels -- a task performed with great enthusiasm.

On June 17, 1772 the *Gaspee* was pursuing a ship off the coast of Rhode Island when she ran aground on a sand bar. Knowing that the schooner was stuck until high tide, the colonials resolved to strike against the hated ship that night. With Whipple as their leader, approximately sixty men in eight boats rowed with muffled oars out to where the *Gaspee* lay stuck. They got to within a few yards before they were spotted. When ordered to identify themselves, Whipple responded, "I am the sheriff of the County of Kent! I have a warrant to apprehend you! So surrender, God damn you!". When the British resisted, shots were fired and the British commander was wounded. He was the only casualty, but the crew was removed and the *Gaspee* was burned.

This event, which occurred a year and a half before the Boston Tea Party, had tremendous repercussions. The King's attorney stated that the incident had "five times the magnitude of the Stamp Act" and Secretary for the Colonies Lord Dartmouth offered a 500 pound reward for the burners and a 1000 pound reward for their leader. Complete amnesty was offered for anyone who would betray their fellow conspirators. There were no takers.

It wasn't until the beginning of the Revolution three years later that word began to leak out about the leaders of the expedition. At this time, British naval Captain James Wallace sent this haughty threat to Whipple: "Sir, on June 17, 1772 you burned her Majesty's vessel *Gaspee* to the waterline and I will hang you from the yardarm." Whipple's reply was full of American insolence. "Sir," he wrote, "Always catch a man before you hang him."

By this time, the battles at Lexington and Concord had been fought and Rhode Island was in the process of organizing its own navy. The free-thinking residents of the smallest colony declared their independence two months before the Declaration of Independence, and their efforts caused one British official to huff, "Rhode Island's government (if it deserves that name) is a downright democracy." On June 15, 1775, the same day as the Battle of Bunker Hill, the Rhode Island legislature authorized formation of its own navy and placed Whipple in command of the flagship. Now that he was a patriot and not a pirate, Whipple wasted no time, and that very day fired on the British frigate

Rose and captured her tender. It was the first official naval action of the war.

In 1776, Continental Congress got around to organizing an American Navy. The fledgling U.S. fleet was authorized to have eight ships and Whipple was named Captain of the *Columbus*, the next to largest ship. His friend John Paul Jones was named first lieutenant on the largest vessel. The new navy faced a difficult task as the British Navy was known to be the world's strongest. A strong blockade threatened to keep American shipping bottled up.

In 1778, Whipple, now commanding the twenty eight gun frigate *Providence*, was ordered to France. Among his cargo were valuable dispatches to the American commissioners there who were in the process of enlisting the French as allies against the British. On the dark and stormy night of April 30, Whipple maneuvered his ship through the Rhode Island harbor that he knew so well. He approached the forty gun *Lark* and could have sneaked past her, but Whipple didn't work that way. Instructing his crew to "pass her on the Narragansett side", he leveled a broadside at the *Lark* and sped on past other British ships and out to sea.

The *Providence* made it to the French port of Nantes in twenty six days. According to some accounts, Whipple took the documents to Paris and was presented to the King of France by Benjamin Franklin. However, it appears more likely that his captain of marines performed this duty and Whipple remained in port with his crew. He did receive his new orders from American Commissioners Franklin, John Adams and Arthur Lee. Whipple's command was to be joined by two other ships on the return trip. This gave him the de facto rank of Commodore, although Congress officially gave no naval officer a higher rank than Captain during the Revolution. Upon his safe return, Whipple received a letter from George Washington stating he was "greatly pleased with the gallant circumstances of your passage through a blockaded harbor."

    This sailor who was earning the admiration of the other founding fathers was short and stout, but possessed a great muscular strength. Because of an accident suffered while unloading captured British cannon, he walked with a limp and carried a cane. Whipple had dark gray eyes and a deep voice that, like many a sailor, was capable of profanity when provoked. Like most sea captains of the era, he was a strict disciplinarian, but he was fair and went to great lengths for his men. It was said he could be surly when things were going well, but that he grew more cheerful as the danger increased. So the American Revolution must have been a happy time for him.

    In 1779, Whipple pulled off one of the most daring feats in the annals of the U.S. Navy. Sailing off the coast of Newfoundland, he came across the Jamaica Fleet -- 150 British ships traveling together. Using a common trick, he hoisted British colors and fell in with the fleet. He initiated a signal conversation with the captain of a straggler ship and invited him over for conversation and a meal featuring fresh meat -- an irresistible invitation. When the British Captain arrived, he was informed that his ship was to be an American prize, and Whipple sent some of his own men over to make up a crew for it. From this ship, Whipple obtained British code books that would enable him to lure other boats away.

While most captains would have been content to swipe a quick prize and run off, Whipple had not yet begun to swipe. He also found out that the British flagship used a lantern as a beacon, so he hoisted one as a lure. In a matter of days, Whipple captured ten ships out of the enemy's largest fleet. And he would have stayed to take more, but he was running out of sailors to make skeleton crews for all his prizes.

When the citizens of Boston saw all the sails offshore they feared the worst, since no American fleet that large had ever been assembled. They were overjoyed to discover that it was Whipple returning with eight of his prizes intact. In addition to capturing 113 cannon, the prizes were valued at about one million dollars, making this the largest haul of the war. With his share of the prize money, Whipple bought a house in Providence and a farm in nearby Cranston.

In 1780 Whipple was ordered south in command of a four-ship flotilla to assist General Benjamin Lincoln in the defense of Charleston. Once there, his command was bottled up in the harbor by a large British fleet and they were forced to serve the campaign on land, where Whipple's luck did not hold. When Lincoln surrendered the city on May 10, 1780, Whipple and his men became prisoners of war.

They were interned in Chester, Pennsylvania, where Whipple spent the next two years. His men were dressed for a southern cruise, so when winter came, Whipple purchased clothing for them from his own pocket. When smallpox broke out, Whipple bought medical supplies and shelter for his men.

He was exchanged for a British officer and returned to Rhode Island in 1782. After the Treaty of Paris guaranteed British recognition of the U.S. government, Whipple had another opportunity for an historic first. In 1784 he was given the honor of sailing the *General Washington* to London--the first time the Stars and Stripes was seen on the Thames.

After this voyage, Whipple returned to his Cranston farm. He served a term in the Rhode Island legislature, but the years after the Revolution meant mainly financial worries for Whipple. Paying for the upkeep on his men at Chester had cost Whipple much of his fortune. When he petitioned Congress for reimbursement and unpaid back wages, payment was authorized but in Continental currency. Due to runaway inflation, this currency was virtually worthless and Whipple soon was in danger of losing his home.

He also didn't get much help from his son-in-law. Whipple's daughter Catherine had married a Revolutionary War officer named Ebenezer Sproat.

Large and genial, Sproat was well-liked but a poor businessman. Whipple gave the couple his Providence home and backed some of Sproat's ventures, which failed and worsened his financial straits. But it was Sproat who first pursued the option of obtaining western lands that were being made available to Revolutionary veterans.

The new nation was cash poor but land rich, and land beyond the Appalachian Mountains substituted for currency. The first organized group to settle in the Northwest Territory was the Ohio Company that settled Marietta. Sproat became involved with this group and was among the original forty eight who landed at the mouth of the Muskingum on April 7, 1788. He was later joined by his family and in-laws, including Whipple's son John. Whipple's other daughter, Polly, married a Dr. Comstock and remained in Rhode Island.

The New Englanders who founded Marietta were an industrious lot. They also had a respect for law and order and a fondness for pageantry. When they had their first court session in September of 1788, an elaborate ceremony was preceded by a parade led by the newly appointed sheriff, Ebenezer Sproat. Friendly Indians, who also enjoyed pomp, saw the 6'4" Sproat and commented something to the effect of "that's one big buckeye there!" in comparing the sheriff to a tall local tree. Their name for Sproat came to be Ha-Hetuck, or Big Buckeye, and eventually all the new settlers were referred to as Buckeyes. After the ceremony, it was discovered that there were no cases on the docket and court was adjourned. The only permanent effect of the first court session in the Northwest Territory was that Ohioans became known as Buckeyes because of Whipple's son-in-law.

    The Native Americans were friendly at first but soon became hostile when they realized the extent of the white man's ambitions. A state of war existed from 1790 to 1795 and during this period settlers rarely dared to venture beyond their cleared fields. When raiding of outposts began, Mariettans believed the British were encouraging the violence and they sent a note to British headquarters in Detroit to try and resolve the issue diplomatically. Chosen to carry this message were John Whipple and Return J. Meigs, Jr., who later became the fourth Governor of Ohio.

    With the help of Indian guides, the pair made it to Detroit safely, but the British commander said it was unsafe to return by the same route. He sent them by ship across Lake Erie to Erie, Pennsylvania, where the overland route was safer. It is not known whether this was young Whipple's first ship voyage or whether it inspired him, but not long after this he left Ohio to pursue a career at sea.

    After the Indians were defeated at Fallen Timbers, it was safe to move out of cramped quarters in town. In 1796, Whipple moved two miles upstream on the Muskingum to a twelve acre farm. Here he labored hard and exhibited no rancor that a former sea warrior should be reduced to dirt farming.

According to one story, Whipple became concerned that local youths were stealing melons from his garden. So one night he sat out with a gun to catch them, but he discovered that it was Indian youths who were doing the pilfering. Whipple did not shoot or sound an alarm but sat undetected and let the raiders take their fill. His attitude towards the original residents was more progressive than most as he understood that the whites were transgressors, so he did not begrudge Indian boys a few melons.

But anyone who thought the old man was mellowing could be in for a rude surprise. Another diarist recorded what happened when Whipple found two soldiers raiding his cucumbers. One soldier saw Whipple and said, "Who are you? If you say anything we will whip you." Whipple responded to this challenge in the same way as he acted at sea, saying, "Well, I must have the first blow." He began to beat the soldiers with his cane, knocking one out before the other surrendered.

As a farmer, Whipple was aware that one of the biggest problems facing the West was how to get produce to market. It was almost impossible to get crops overland through the mountains without spoilage. For this reason, moonshining was an important skill, for a corn crop could be made portable by being made potable. But the lack of roads meant that waterways served as the highways. And west of the Alleghenies all waterways led to New Orleans, then a Spanish possession. The new American nation negotiated the right to traverse the

Mississippi in 1795, but they were still at the mercy of the Spanish. Although keelboats could be poled upstream at an agonizingly slow pace, most river traffic was one way. Laden boats landing in New Orleans could neither go upstream nor downstream and therefore had only one market for their goods.

This would not do for the New Englanders who settled Marietta. Around them were trees that made excellent lumber, hemp growing wild, and a group of people with shipbuilding experience. In the winter of 1800, they resolved to build an ocean-going vessel and sail it downstream to the sea. One more valuable resource they had available was Abraham Whipple to captain the ship.

Some said that Whipple had to be persuaded to take the helm, while others claimed his reluctance was but a pretense and he was delighted and eager. But the result was that the Commodore agreed to lead one more voyage at age sixty-eight and add another first to his maritime exploits. He was not only the obvious choice, he was the only choice, because no one else in town knew anything about navigation.

At about the same time, merchants in Elizabeth, Pennsylvania, on the Monongahela were also building an ocean going vessel. This group had previously built a schooner that was alleged to have made it to the Gulf of Mexico, but there is no confirmation of this. Now they were at work on the *Monongahela Farmer*, a square rigger that could carry cargo to the sea.

The Marietta group was headed by local merchants Dudley Woodbridge Sr. and Jr., Benjamin Ives Gilman, and Charles and Griffin Greene. The last named was a cousin of Nathaniel Greene and a friend of Meriwether Lewis, who visited him in Marietta two years later on his way west to meet up with William Clark. These men employed experienced woodworkers who took full advantage of the area's abundance. The hull was made of black walnut, with red cedar and locust used for the top timbers and gunwales. Long ashwood sweeps were made for when the boat had to be rowed, which was often. The two-master was finished in time for the spring freshet that would raise the river levels. This was necessary because the ship would be carrying 110 tons of pork and flour.

The brig was christened the *St. Clair*, after territorial Governor Arthur St. Clair. This was a political act that showed support for the Governor at a time when he opposed statehood for Ohio. St. Clair, a Federalist, knew that statehood would result in two new Democratic senators and a loss of his job. The Marietta area was one of the few bastions of Federalist support, and some hoped to hold out for creation of two states--a Democratic one headquartered in Cincinnati and a Federalist state with a capital at Marietta.

It is not known the exact day the *St. Clair* was launched, but the owners gave Whipple his final instructions on April 19, 1801. So it must have been right around that date that a crowd gathered to see the ship off. Governor St. Clair was in attendance and the fondness for pageantry was again indulged. Local boat builder and poet Jonathon Devol, whose brother had served with Whipple before being killed in the Revolution, summed up local feeling for Whipple with this tribute:

"The Triton crieth, 'Who cometh now from shore?'
Neptune replieth, 'Tis the old commodore'
Long has it been since I saw him before,
In the year seventy-five from Columbia he came,
The pride of the Briton on ocean to tame:
And often too, with his gallant crew,
Hath he crossed the belt of ocean blue.
On the Gallic coast, I have seen him tost,
While his thundering cannon lulled my waves
 and roused my nymphs from their coral caves;
When he fought for freedom with all his braves,
In the war of the Revolution.
But now he comes from the western woods,
Descending slow with gentle floods,
The pioneer of a mighty train,
Which commerce brings to my domain.
Up, sons of the wave, Greet the noble and brave!
Present your arms unto him.
His gray hair shows, Life nears its close:
Let's pay the honors due him.
Sea-maids attend with lute and lyre,
And bring your conchs, my Triton sons;
In chorus blow to the aged sire,
A welcome to my dominions."

After setting sail, the first landmark seen by the crew of the *St. Clair* would have been Blennerhassett Island, fourteen miles downstream. Here stood the newly completed mansion of English exiles Harman and Margaret Blennerhassett. Whipple certainly knew Blennerhassett, who was a business partner of Dudley Woodbridge. In fact, it was Woodbridge's firm that built the flat boats that Aaron Burr was to use on his ill-fated expedition that caused Blennerhassett's fall five years later.

But the *St. Clair* sailed on into previously unsailed waters. Whipple experimented with ideas such as sailing backwards and dragging anchor to maintain control in the narrow, shallow channel. The shifting course of the river made it difficult to rely on the wind for sailing and Whipple's landlubber crew could be relied upon for only basic maneuvers. Still, the *St. Clair* made good time, arriving at Cincinnati on April 27.

Here they were greeted by an enthusiastic crowd that knew of their mission. The newspaper *The Western Spy* reported of the event that "on her arrival the banks were crowded with people, all eager to view this pleasing presage of the future greatness of our infant country. This is the first vessel which has descended the Ohio equipped for sea."

The biggest obstacle facing any boat on the Ohio River was the Falls of the Ohio at Louisville. Navigable only at high water, the Falls claimed many a boat, but Whipple was able to sail the *St. Clair* through safely. Not all future captains fared as well, as three Marietta-built ships were dashed on the rocks here in the next six years.

The Falls also ended any hope the *Monongahela Farmer* had of competing with the *St. Clair*. The Pennsylvania ship had the disadvantage of starting much farther upstream. She was not launched until April 23 and did not even make Pittsburgh until May 13. By the time they got to Louisville the water was so low that they were forced to wait over three months. When they finally did make it to New Orleans, their cargo of flour had spoiled and could only be sold to the cracker manufacturers.

Whipple and his crew pushed on, apparently encountering enough dangers to keep the Commodore from getting too surly. From Kentucky he wrote to the ship's owners that "we got once in the trees by a sudden shift of the wind but soon got off by carying the ancar astarn (sic)." Of his crew, Whipple conceded "my people behave wail(sic) according to what they know as seamen."

In addition to the dangers posed by an unknown river, Whipple also had to worry about both Indians and pirates. The Ohio area tribes were at peace, but little was known about some of the downstream tribes. And river pirates were a notorious threat on the western rivers at this time. By 1801, the cutthroats who operated out of Cave-in-Rock in Illinois had recently moved their operations towards Natchez on the Mississippi. But apparently neither Indians nor pirates chose to tangle with the brig, and the *St. Clair* reached New Orleans in June.

They had covered 2,000 miles of river in six weeks, an accomplishment that greatly pleased the captain. "Commodore Whipple thinks it the greatest thing he ever did," wrote one correspondent.

In New Orleans, the Spanish were free to charge exorbitant fees for the privilege of docking. But the *St. Clair* had other options available. Whipple refused to pay the fees and anchored in mid-stream, stopping only long enough to take on supplies. Then he took off for the Gulf of Mexico, where he was able to smell the salt water for the first time in seventeen years. The *St. Clair* then made for Havana, making much better time on the open sea than on the cramped river.

In Havana, another Spanish possession, port authorities charged a duty of $20 a barrel. But Whipple's welcome pork and flour easily covered that, selling for $40 a barrel. With the profits he purchased a cargo of sugar to take back to the U.S.. And in an amazing stroke of good fortune, they ran into John Whipple in port. Better yet, he was at liberty to join the crew for the return voyage. This gave the crew an additional navigator and lessened their dependence on the Commodore.

On the return voyage, the crew of the *St. Clair* faced their greatest crisis. Yellow fever broke out on board and soon took a serious toll. The Whipples had been exposed to this tropical threat before and were not affected, but nearly a third of the crew died. Whipple was able to guide the survivors home after the disease had run its course and the *St. Clair* reached Philadelphia in late summer.

Here the Commodore not only sold his sugar under favorable terms, but he sold the ship for a good price as well. After staying in Philadelphia for a few weeks, Whipple and crew walked across Pennsylvania and came back to Marietta with enough money for farmers, boat builders and merchants to share. Then, like Cincinnatus before him, he returned to his plow.

Whipple never again returned to the sea, but his last voyage spawned an industry that saved the economy of the nascent town. In the next eight years, 26 ocean going ships were built in Marietta, and the city was named a U.S. Port of Entry. One of the Marietta ships was nearly confiscated in St. Petersburg, Russia, because officials had never heard of the port of Marietta and could not believe it was so far from the ocean. Zadock Cramer, whose guidebook *The Navigator* was the early bible of Ohio River travel, said of Marietta in his 1807 edition, "shipbuilding is carried on here with spirit."

Jefferson's Embargo Act of 1808 restricted commerce and brought a sudden end to the inland shipbuilding industry. A few years later, steamboats started plying the western waters, and their ability to come back upstream made them the profitable boats to build. There was brief revival of shipbuilding in the 1840's, but by then it was easier to tow the ships to sea by steamboat before ever unfurling a sail.

As for Whipple's immediate compensation, he had to sue to get his wages. The New Englanders were fond of litigation and this was apparently how they settled many business deals. Whipple filed suit against the owners for wages, commission and boarding in Philadelphia. He even filed a separate suit for reimbursement for twenty four gallons of peach brandy he'd purchased for the crew in Havana. For his efforts, he was awarded a settlement of $478.17 1/2.

But Whipple's financial troubles were ended in 1811 when Congress finally granted him a pension for his Revolutionary War service. This enabled him and Sarah to live out their final years in relative comfort. Sarah died in October, 1818, and Abraham Whipple followed on May 29, 1819. He was buried in Marietta's prestigious Mound Cemetery, the final resting place of many Revolutionary veterans, with the following inscription:

SACRED
TO THE MEMORY OF
Commodore Abraham Whipple,
WHOSE NAME, SKILL, AND COURAGE,
WILL EVER REMAIN THE PRIDE AND
BOAST OF HIS COUNTRY.
IN THE LATE REVOLUTION, HE WAS THE
FIRST ON THE SEAS TO HURL DEFIANCE
AT PROUD BRITAIN;
GALLANTLY LEADING THE WAY TO
ARREST FROM THE MISTRESS OF
THE OCEAN, HER SCEPTER,

AND THERE TO WAVE THE
STAR-SPANGLED BANNER.
HE ALSO CONDUCTED TO SEA,
THE FIRST SQUARE-RIGGED
VESSEL EVER BUILT ON THE OHIO,
OPENING TO COMMERCE
RESOURCES BEYOND CALCULATION.

CHAPTER VIII
William Maxwell and Nathaniel Willis: Pioneer Printers

    Settlement came slowly in the first years of the Northwest Territory because of the Indian wars. However, newcomers still arrived at the already established posts, and they gradually brought with them the trappings of eastern civilization. Among the most important of these was the printing press. The advent of printing in a new area conferred a certain status and sense of permanence. And for Americans the printed word has always been a crucial component of the settlement process.

The first newspaper in America was *Publick Occurrences* , which was printed in Boston in 1690. However, it was suppressed after only one issue, and no one made another attempt until 1704. So newspapers in this country really began at the beginning of the 18th century. After playing an important role in colonial life and the movement for independence, the press was moving west by the end of the century.

It was in 1786 that printing first crossed the Alleghenies. In that year the *Pittsburgh Gazette* began operation. The following year the *Kentucky Gazette* was started in Lexington. This gave Kentucky's statehood efforts a boost since until then citizens had been trying to organize a statehood or secession movement without the benefit of a press to get the word out.

In 1793, printing crossed over into Ohio, and William Maxwell was the future state's Gutenberg. Maxwell was born in New Jersey in 1755, the son of Scottish immigrants. He served in the Revolution but it is not known where he learned the printer's trade. Maxwell came west after the war and set up a printing office in Kentucky. He did some work there, but another printer was already established in Lexington and there was not enough work to support two presses.

So Maxwell crossed the Ohio River, where he had the field all to himself. He moved his crude printing equipment to Cincinnati and set up shop in a log cabin at the corner of Front and Sycamore Streets. Here, on November 9, 1793, he printed the first issue of the *Centinel of the Northwest Territory.*

The first newspaper in what is now Ohio was quite different from newspapers of today. For one thing, it was only four pages, with three columns on each 8 1/2 by 10 1/2 inch page. There were no pictures, very little local news other than accounts of recent Indian raids, and very old national news. There were reprinted stories from London dated July 15, and stories from New York and Philadelphia dated September 4. The comic section of this first issue consisted of this anecdote: "Milton was asked by a friend whether he would instruct his daughters in the different languages. To which he replied, 'No sir, *one tongue* is sufficient for a woman.'"

On his masthead, editor Maxwell put the slogan "open to all parties, influenced by none". He explained his choice of name for his newspaper by saying that he hoped to serve as a sentinel, by observing, warning and helping to bring about a common defense. He also apologized for mislaying his subscription list and asked subscribers to come to his office to pick up their first weekly issue.

If there were any ancillary occupations to being a pioneer printer, they would have to be postmaster and politician. There were no wire services with instant news -- the western printer was dependent upon eastern newspapers that came by post. For this reason postal routes were essential to the printer and on more than one occasion late mail caused a frontier printer to miss a deadline.

The political nature of most news meant that pioneer printers were also involved in politics, often in an openly partisan role. Joseph Carpenter, a rival who started his own Cincinnati newspaper in 1799, observed that "fully three-fourths of the printers in the United States have attached themselves to one party or the other, and their pages have been devoted to justifying their own and ridiculing the other." Media bias was more open in those days and printers often found themselves appointed to political posts when their party was in power.

Another factor that brought politics and printing together was that government was one of the few regular customers of the printer. Maxwell was appointed printer to the legislature and soon had enough work to keep his press rolling. It was no coincidence that the three towns that had presses during the territorial period -- Cincinnati, Chillicothe and Marietta -- were the three towns that fought to be named capital.

The Ordinance of 1787 had said nothing about capitals. The "legislature" during this early period consisted of Governor Arthur St. Clair, Territorial Secretary Winthrop Sargent, and three appointed judges, who actually served more as legislators. The first "capital" of the Northwest Territory was a blockhouse of the Campus Martius stockade that the Mariettans provided for St. Clair after his arrival on July 9, 1788.

The first three judges appointed were all from the Ohio Company, but Samuel Parsons drowned in 1789 and James Varnum died in 1792. They were replaced by John Symmes and George Turner of the Symmes purchase settlements in the Cincinnati area. Meetings of the legislature were wherever a quorum could best be assembled.

The Territorial Secretary had gubernatorial power whenever St. Clair was absent from the territory. This was quite often, since St. Clair was needed for negotiations with both Indian tribes and federal officials. Sargent was a stern, aristocratic New Englander whose main service to the people was that he made St. Clair look good by comparison. During one absence, Sargent made rules in the Cincinnati area that restricted alcohol, gambling and firearms -- three of the favorite pastimes of the frontiersmen.

Maxwell took an editorial stand against Sargent's actions and incurred the wrath of the Secretary. So it was with glee that Maxwell announced the return of St. Clair to the territory, saying "Happy period -- at which tyranny and despotism must once more lay down the arm of cruelty and oppression. Let the patriotic lovers of rational liberty again rejoice."

When Maxwell came to Cincinnati he was a bachelor, but not long after he arrived he met Nancy Robins. She was a woman with an impressive pioneer pedigree, having been one of the first settlers at Wheeling. In fact, she might have been as famous as her friend Betty Zane if she had not been such a good bullet maker.

In September of 1782, Wheeling's Fort Henry was attacked by a group of Indians and Tories in one of the last battles of the Revolution. Nancy's father was killed in the initial surprise assault, and she barely made it to the fort herself. Those inside were trapped without an adequate supply of gunpowder and it was decided that a woman running to a nearby cabin for more would be less likely to draw immediate gunfire. Nancy was considered too valuable as a bullet moulder, so instead young Betty Zane was picked and became the heroine who saved the fort. Nancy's widowed mother later married Betty's brother Ebenezer Zane.

After a brief courtship, William Maxwell and Nancy Robins were married and they went on to have eight children. Nancy helped with her husband's business, even though she was apparently illiterate. In 1796, she did the binding of the first book that was published in what is now Ohio. The 225 page book was sewn with waxed ends tipped with bristles and was titled *Laws of the Territory of the United States Northwest of the Ohio*, but was generally called Maxwell's Code.

In 1794, Maxwell was named to replace the original postmaster of the city of Cincinnati. This job gradually became full time, and in 1796 he sold his newspaper and press to Edmund Freeman, who continued the paper under the name *Freeman's Journal.* Maxwell moved to Dayton in 1799 and later bought a farm along the Little Miami in Greene County. He was elected to the first General Assembly of the State of Ohio, thus completing the career trio of printer, postmaster and politician. He also served as judge and sheriff of Greene County before dying there in 1809. Nancy remarried and had more children and moved to Illinois, where she lived to be 108.

Freeman succeeded Maxwell as printer to the Northwest Territory, since his press was still the only one in the territory. However, competition arrived in 1799 when Joseph Carpenter started *The Western Spy and Hamilton Gazette*. Carpenter, a Massachusetts native, promised fast news and objective reporting. His four page 12 inch by 19 inch weekly was sold to city subscribers for $2.50 a year, but the price shot up to $3 a year for out of town delivery. Carpenter apparently meant his timeliness promise, for , on January 7, 1800 he reported the death of George Washington, a mere 24 days after the event occurred.

In 1800, Freeman moved his operations to Chillicothe, but it wasn't competition that drove him there. His move was because the Territorial capital was moved there and he needed to follow the political news as well maintain his duties as printer to the legislature. The Territory now had the 5,000 voters necessary to elect a legislature, and at the first session they voted to move to Chillicothe for the next session.

But Freeman did not last long at his new location. He was already suffering from yellow fever, a disease that finally claimed him on October 25, 1800. Shortly before his death, he sold his press and equipment to Nathaniel Willis, who started a new paper called *The Scioto Gazette and Chillicothe Advertiser.*

Willis had a solid printing background and impressive patriotic credentials. Born in Boston in 1755, as a teenager he participated in the Boston Tea Party. He was involved with the militant Boston newspaper the *Independent Chronicle* from 1776 to 1784, and during part of this period he served as a staff officer of General John Sullivan of the Continental Army.

After the Revolution, Willis took his printing equipment to Winchester, Virginia, where he founded a newspaper. He also later founded papers in nearby Shepherdstown and Martinsburg. While he wasn't the first publisher in Ohio, his Shepherdstown paper was the first in what is now West Virginia. It was here that Willis met Thomas Worthington, who at that time was urging people to move west to the Ohio country.

It is not known exactly when Willis arrived in Chillicothe, but he started his newspaper in the fall of 1800. His financial backer was Winn Winship, a businessman recently arrived from Maryland. Their office was located on Second Street in a house that Worthington had built in 1798 that featured the first glass windows in town.

In his first issue on October 10, Willis advertised for two journeyman printers and promised aggressiveness in the pursuit of news. He boasted of having made arrangements with "correspondents in the Atlantic States as to be furnished in the course of two weeks with the latest papers printed in New York, Philadelphia, Baltimore, City of Washington."

This proved to be a difficult promise to keep, and Willis sometimes had to change his publication dates to suit the mail's arrival. In one issue he had to apologize that "the Eastern post-rider arrived last evening without any mail; we therefore have been disappointed in giving our readers that important information we anticipated." This situation got so bad that Willis tried to take matters in his own hands, advertising for a post rider to go weekly from Chillicothe to Cincinnati and deliver mail as well as subscriptions.

The dependence of the printer on the post rider was made worse by the deplorable state of roads in the area. In fact, in the entire territory there was only one road, and it was built by Ebenezer Zane, the stepfather of Nancy Robins Maxwell. In 1796, Zane petitioned Congress to build a road from Wheeling to Limestone (now Maysville), Kentucky. This road, which was really not much more than a path, provided an overland route to Kentucky. It served as a sort of Ohio version of the more famous Natchez Trace, the route that Mississippi River flatboatmen would use when returning upstream. Also, since the Ohio bcould be ice-choked in winter and too shallow in summer, it gave downstream travelers an option as well.

For payment, Zane asked for land grants and the right to establish ferry service where his Trace crossed the Muskingum, Hocking and Scioto Rivers. This turned out to be a very shrewd proposal, as the towns of Zanesville, Lancaster and Chillicothe came to be located on or near the grants provided by the building of Zane's Trace.

Another problem that Willis had to deal with was a lack of paper. Chillicothe is known today for paper mills, but the first of these did not arrive until 1810. Until then, a shipment of paper from Pennsylvania took two weeks to arrive and was of poor quality. Of course, so was the worn typeface and crude woodcut engravings that Willis used in his ads and news stories.

Despite these conditions, Willis and his paper both prospered. He was able to buy out Winship and purchase several lots in town. The *Scioto Gazette* grew in influence, especially in the fight for Ohio statehood, where Willis took a decidedly partisan view.

The Scioto Valley was settled mainly by Virginians, many of whom arrived by way of Kentucky. Men like Worthington, his brother-in-law Edward Tiffin, and Nathaniel Massie came here with political leanings towards the Democratic-Republican party of Thomas Jefferson. The Miami River settlements were similarly inclined, but the residents of the Muskingum River Valley leaned towards the Federalist Party of Washington and St. Clair.

The Ordinance of 1787 stipulated that statehood could be achieved when 60,000 voters inhabited the Territory. It also stated that a minimum of three and maximum of five states were to be formed from the Northwest Territory. What was not stated was how these states were to be divided up, so no one knew in what parts of the territory the 60,000 voters had to be located.

In the first period of settlement, St. Clair had absolute veto power and only three judges to deal with. But when settlement totaled 5,000 voters, a representative legislature could be elected, which required more compromise. St. Clair discovered his new limitations right away when the first legislature chose William Henry Harrison to be the Territorial representative to Congress instead of his candidate, Arthur St. Clair Jr. The Territorial representative was a non-voting delegate, and Congress was generally too preoccupied to give the west much attention, which worked to St. Clair's advantage.

But effective lobbying by Harrison and Worthington gradually changed this, and after 1800 the new Jefferson administration was more receptive to westerner's concerns. In order to stave off statehood and preserve his job, St. Clair needed to use his own political skills and isolate the Chillicothe Democratic contingent that Willis was a vocal part of.

St. Clair approached the Cincinnati representatives with a proposal that would divide the territory into three states with capitals at Marietta, Cincinnati and Vincennes (Indiana). He reminded them that if the state border were located near the Miami River, then centrally located Chillicothe would be the logical capital and Cincinnati would be shut out. So with this gerrymandering proposal St. Clair aligned the Federalists at Marietta with the Democrats at Cincinnati. In doing so, he hoped to isolate the Chillicothe region and guarantee that it would be years before any of the regions would have enough voters for statehood.

St. Clair's plan was defeated by the federal government, which responded to the behind the scenes efforts of Harrison and Worthington. In 1800, Congress passed the Land Act, which resolved Connecticut's claims on the Western Reserve and settled most of the boundaries of what is now Ohio. The remainder of the former Northwest Territory was renamed Indiana Territory, with Harrison named Governor and John Gibson as Secretary.

In 1802, Congress passed the Enabling Act, which lessened the requirements for voter eligibility and greatly increased the already increasing number of voters--particularly Democratic ones. This made statehood a certainty, and when St. Clair openly denounced the bill, Jefferson had cause to remove him.

For Willis, this final process brought his journalistic career full circle. In 1776 his paper had opposed the British and helped bring about a new nation and now his efforts were helping to bring a new state into that nation. In the December 18, 1802 edition of the *Scioto Gazette* the phrase "State of Ohio" replaced "North-Western Territory" on the masthead. The first elected General Assembly of Ohio did not convene until the following March, but members had already been elected and the battle for statehood won. It is an indication of how fast things moved on the frontier that Edward Tiffin was elected the first Governor of Ohio a mere five years after moving here, and Chillicothe became the capital of a new state just seven years after being founded.

A few years later Willis ended his 30 year journalistic career when he sold the *Scioto Gazette* in December, 1805. When his successor ended the old English custom of using the letter "f" for "s" in the lower case, it symbolized the end of an era in pioneer printing.

Willis didn't retire, as he remained concerned with postal service. He became involved with the first stagecoach operation in Ohio, which delivered mail along Zane's Trace from St. Clairsville to Kentucky. He moved to a farm along the Trace in Pike County, where he also operated a tavern. He died there in 1831.

Among the children he and his wife Lucy had was a son, also named Nathaniel. He learned the printing trade from his father in Virginia, but had gone back to Boston in 1796 rather than come to Ohio. He published primarily religious material in the east, and his son, Nathaniel Parker Willis, became a journalist and poet and one of the first Americans to make a full time living as a writer. While his son and grandson remained back east, it was Nathaniel Willis who came west after helping to gain freedom for his country to assist in the establishment of his adopted state.

CHAPTER IX:
Benjamin Tappan:
The Curmudgeonly Politician

As the 18th century came to a close, settlers were flocking to the Ohio country in ever increasing numbers. When Connecticut tried to develop its claims to northeastern Ohio, a whole new region was opened for settlement. Of the many New Englanders who came west in search of new opportunities, one of the most unique was Benjamin Tappan.

Arriving from Massachusetts in 1799, Tappan founded the city of Ravenna and was the first settler in what is now Portage County. But he is intriguing for more than just his pioneer pedigree -- Tappan was a man with talents and interests as diverse as those of his great-great-uncle Benjamin Franklin. He was a painter who studied under the noted portraitist Gilbert Stuart and he loaned a young Thomas Cole his first set of brushes. He was a jurist who wrote the first book of legal review published in Ohio and was the law partner of Edwin Stanton. He was interested in educational reform and also served as chairman of the state canal commission that revolutionized frontier transportation. He was a scientist who collected and wrote about minerals, was founder and president of the Historical and Philosophical Society of Ohio and played a role in the establishment of the Smithsonian Institution. And he was a U.S. Senator who was elected despite a curmudgeonly honesty that regularly cost him votes.

A rival newspaper once called Tappan "the most unpopular, the ugliest and the smartest man of his party in this part of the state." He was a man who didn't suffer fools gladly and he saw fools everywhere. This view often got him into trouble, such as when he was successfully sued for slander by his brother-in-law. He was a radical member of the Democratic Party that supported the common man despite his own patrician bearing. He favored abolition of slavery, yet was a freethinker who ridiculed organized religion. Tappan's whole life could be seen as a solitary revolt against his upbringing, yet he was precisely the kind of man needed to bring civilization to Ohio.

Benjamin Tappan was born in Northampton, Massachusetts on May 25, 1773. He was the second of ten children and the oldest of six sons born to Benjamin and Sara Tappan. His father was a goldsmith who as an apprentice had married his master's daughter. Sara Homes Tappan was the granddaughter of Benjamin Franklin's sister Mary, but her branch of the family did not maintain Franklin's liberal religious views. Sara raised all her children to be strict orthodox Calvinists, and it was appropriate that for a while the family lived in the former home of Jonathan Edwards, the most famous hellfire and damnation preacher of New England Calvinism.

Most of the children followed the family line in matters of religion. The youngest son Lewis shocked the family with a youthful flirtation with Unitarianism, but he soon returned to the fold with the passion of a convert. But Benjamin rejected his parents' religion with a vehemence that was immediate and permanent. Though he always conducted himself with rectitude and probity, he made sure it was known that religious doctrine was not the reason for his behavior.

With family roots in the colony going back to 1637, the Tappans came from stock that was talented and respected, but not wealthy. The family planned to send their eldest son to Harvard, but when

Benjamin realized what a hardship this would be for his parents he announced that he would prefer to learn a trade. From the ages of 14 to 21 he worked at several mechanical trades, among them goldsmithing, copper plate engraving, and the making of both musical instruments and clocks. He also read widely, took at least one sea voyage, and generally got as good an education as college could have given him.

At age 21, the artistically inclined Tappan resolved to go to Europe and become a painter. To learn this trade he went to New York City, where he studied for six months under Gilbert Stuart, the most famous of early American portrait painters. Stuart had recently returned from Europe, and it was about this time that he did his most famous portrait of George Washington that is on the one dollar bill. Though Tappan enjoyed the cultural opportunities offered in New York, his circumstances forced him to live frugally, and in 1796 he decided to become a lawyer.

He studied law for the next three years under the tutelage of Gideon Granger of Suffield, Connecticut. It was here that Tappan came to reject the politics of his parents in addition to their religion. The New England Calvinists tended to be conservative Federalists but Granger was a prominent member of the Jeffersonian party. In fact, Jefferson named Granger his postmaster general after being elected President in 1800. This was a period of deep partisan turmoil and Tappan eagerly joined the more liberal Jeffersonians and used his mocking wit to write satirical verses aimed at the Federalists. Tappan was admitted to the bar in 1799, but he never practiced law in Connecticut because soon afterwards he left for Ohio.

Tappan's father had speculated in lands claimed by Connecticut in the Northwest Territory. His revolt against his family was never so complete that he was estranged from them, and he and his father became partners in the development of the western lands. The son's role was to go west to explore the land and make it ready for profitable resale.

The original Connecticut colonial charter specified no western boundary. This was true of several other colonies as well and led to considerable confusion during westward expansion. During the Revolution, the settlers of the Wyoming Valley in central Pennsylvania considered themselves residents of Connecticut. Even after this issue was settled, the state still claimed a 120 mile strip of land west of Pennsylvania as a western reserve, but it wasn't until after the Greenville Treaty that this land was safe for settlement. The Connecticut legislature then set up the Connecticut Land Company, which was authorized to sell the land unsurveyed. The westernmost 25 mile strip of the Reserve was called the Firelands, and this land was to be used to compensate residents of coastal towns that had burned by the British during the Revolution. Connecticut did not fully relinquish her western claims until 1800, when the area finally came under the jurisdiction of the Northwest Territory.

In 1796 a surveying party under the leadership of Moses Cleaveland entered the Western Reserve. Despite much hardship in the primitive backwoods, they surveyed much of the Reserve and laid out a town at the mouth of the Cuyahoga that they named after their leader. Moses Cleaveland never returned to Ohio but the city named for him thrived, albeit with a slight spelling change.

The survey crew continued their work the following year, working in uninhabited land with no roads and widely varying terrain. The title holders back in Connecticut in the meantime had no idea what sort of land they had purchased. The elder Benjamin Tappan's shares included 14 separate parcels, the bulk of which constituted two-thirds of what is now Ravenna Township in Portage County. The elder Tappan was not normally so speculative, and he was relieved that his eldest son was willing to enter this venture with him. The younger Tappan was also an eager partner, for despite his confident exterior, he was uncertain he could make a living as a lawyer in Connecticut, where the bar was filled with capable attorneys.

On April 19, 1799, the 25-year-old Tappan left Northampton for the Ohio country. He took with him a yoke of oxen, a cow, farming equipment and some hired men, including a blacksmith named Kellog and his family. The livestock he sent overland with a relative, while Tappan and the rest of the party took a water route through New York, traveling up the Mohawk River and crossing over to Lake Ontario. Here they met David Hudson, another landholder en route to the Western Reserve, who traveled with them the rest of the way.

While on the Niagara River between Lake Ontario and Lake Erie, they were caught in strong currents and barely made it to shore before they would have been swept over Niagara Falls. Townspeople had watched Tappan proceed without warning him and after surviving what he called "the most perilous adventure of my life", he went back and told them what he thought of their conduct. It is a measure of how deeply Tappan could hold a grudge that 40 years after this incident he wrote that "I was really gratified that the whole village was destroyed in the War of 1812."

On Lake Erie the party was delayed by late season ice before they finally made it to Cleveland. From here they proceeded upstream on the Cuyahoga to the site of the now-abandoned village of Boston, where the river became too shallow. Tappan then went overland along the surveyor's township boundary lines until he arrived at his father's land on June 10. Other than a man named Honey, who had a cabin near Mantua, he was the first settler in what is now Portage County.

Almost immediately the projected settlement ran into problems. Tappan had left Kellog guarding most of the supplies on the Cuyahoga, but Hudson had lured the blacksmith away with an offer of 200 acres if he would work for him. Tappan returned to find his provisions had been abandoned and plundered. Then, when he hurried to haul the remainder overland, one of his oxen became sick and died.

Literally down to his last dollar, Tappan sent one employee to Erie, Pennsylvania to obtain a line of credit from his father, while he searched for a replacement ox. He walked the township lines until he came to a road from Warren to Youngstown, and at the latter place was able to obtain an ox at a fair price and on credit. But by now it was too late in the year to raise a decent crop.

A religious man, Hudson later came to Tappan and apologized for luring Kellog into breaking his contract, but the two continued to have an uneasy relationship. Hudson did invite Tappan to deliver the Fourth of July speech at his settlement in 1801, but then Tappan proceeded to alienate the crowd with a speech "interlaced with many grossly illiberal remarks about Christians and Christianity," according to one minister present. Hudson became a leading citizen of the area and one of the founders of Western Reserve College, but in his accounts of early settlement he barely mentioned Tappan. For his part, Tappan did grudgingly concede that "it is rare that men of such pretension to sanctity are honest, but Hudson was an exception, notwithstanding his conduct to Kellog and some other aberrations."

Tappan continued to clear land with his small crew, though for a three-week period he was the sole resident of his township. Eventually a 22' by 18' cabin was completed, and Tappan moved in and began a new era appropriately enough on January 1, 1800. The first winter was a harsh one. When Tappan ran out of flour in February with 20 inches of snow on the ground, it was a 45 mile trip to the nearest mill.

Tappan decided to call his town Ravenna, either because a brother had visited that city in Italy or because he liked the name. He could have made his town eponymous, as did Cleaveland and Hudson. This was a common Western Reserve tradition, as among the early pioneers who named towns after themselves were men named Kirtland, Kinsman, Stow, Hubbard, Wadsworth, Tallmadge, Struthers and Hinckley. But to do this would have meant following the crowd, which was something Tappan never did.

The following June the pioneer lawyer had his first case when he assisted in the defense of two settlers accused of murdering two Indian men and a child. Tappan admitted to being nervous, but it was rare that an acquittal could not be obtained for whites killing Indians, and this, the first murder trial in the Western Reserve, was no exception. With few other interruptions, Tappan and his crew cleared much new land in 1800. In December of that year, he returned east to spend the winter. When he came back in the spring, he was accompanied by a new bride, the former Nancy Wright, whom he had courted over the winter.

Tappan soon became involved in territorial politics, where he advocated statehood and opposed the Federalist governor, Arthur St. Clair. He accused the administration of corruption, saying that St. Clair charged fees for licensing every activity because he "required more than his salary to supply him with drink." Yet he did marvel at the Governor's capacity, noting that at one reception St. Clair met 24 different men and had a brandy with each of them and still remained steady.

When statehood became a reality, Tappan ran for the state senate seat for Trumbull County, which then covered the entire Western Reserve. However, confusion about the date of the election, which Tappan claimed was caused by his opponent, led to a small turnout and the election of Samuel Huntington to the Senate. But when Huntington was subsequently named to the Ohio Supreme Court, Tappan won the special election to replace him and took a seat in the Ohio Senate on December 1, 1803. It seemed a promising beginning for the 30-year-old Tappan, yet it would be 35 years before he was again elected to office.

He was defeated in his bid for reelection in his strongly Federalist district. But he did score a political coup when Ravenna was named the seat of newly created Portage County in 1808. It was a land speculator's dream to be in the center of a new county, since land values would go up at the seat of government. And not only were the lines drawn to put Ravenna in the center of the new county, but Tappan's house was designated the first county court house.

Tappan had bought his father out in 1806, so this latest development was a real financial windfall for him. Yet the town still grew slowly and Tappan considered it a backwater without sufficient legal work to keep him busy. He looked about for greener pastures and in the spring of 1809, he and Nancy moved to Steubenville, where his legal practice could flourish and the populace was more receptive to his Jeffersonian politics.

He was living here when the War of 1812 began. Tappan had been named a major in the Ohio Militia by his friend Elijah Wadsworth, who held the rank of major general. When war was declared, Tappan was ordered to gather the Jefferson and Harrison County militia and march to Cleveland. The news that the American force had surrendered at Detroit filled the state with panic and Cleveland was rumored to be the next target of the British. Tappan had no military experience, but he read manuals on drilling and was at least able to train his amateur troops somewhat. He soon found, however, that the problem of supplies was the first issue faced by any army. He also found that politics played a large role in military matters. When Wadsworth resigned after a dispute over supplies with General William Henry Harrison, Tappan also resigned his commission.

He returned to Steubenville, where for the rest of his life he was devoted to community and cultural affairs that helped advance his adopted state. In 1816 the legislature appointed him President of the Fifth Circuit Common Pleas Court. Here he served enthusiastically for a seven-year term that ended when he unsuccessfully ran for State Supreme Court Justice. Tappan kept notes of the court's rulings and eventually published a book entitled *Reports of Cases_Decided in the Courts of Common Pleas of the Fifth Circuit of Ohio*. This book, which came to be known as "Tappan's Reports", was the first book of legal review published in Ohio, and was often cited. It was printed by James Wilson, grandfather of Woodrow Wilson. Wilson had been working for the *Aurora*, a Philadelphia newspaper that was the leading Jeffersonian publication of its day, when Tappan lured him west to join the *Steubenville Western Star*.

Tappan was active in other civic and cultural affairs as well. In 1822 he was named to the state canal commission, and he became chairman the next year when Thomas Worthington left the board. This panel was responsible for overseeing the development of the canal system that revolutionized transportation in the frontier state. The field work involved also enabled Tappan to indulge his passion for geology. He became a known expert in minerals and shells and their classification. He collected specimens and wrote articles that were published by the Historical and Philosophical Society of Ohio, a scientific group of which he was both founder and president.

Tappan also continued to be interested in art. He was the first to offer encouragement to Thomas Cole, a Steubenville teenager who went on to become America's foremost landscape painter and the founder of the Hudson River School. Cole was a dreamy English immigrant when he arrived in Ohio with his family in 1819. When he became interested in painting, it was Tappan who loaned him his first set of brushes. Cole tried to become an itinerant portrait painter in 1822, but after unsuccessful stops in St. Clairsville, Zanesville and Chillicothe, he returned to Steubenville in poverty. Cole's first landscapes were theater backdrops for a local thespian troupe, but when vandals destroyed the equipment he had borrowed from Tappan, he gave up and left the state. Three years later he became a sensation in New York City, and soon became the most famous American painter of his era.

Another area where Tappan had an interest was educational reform. He became interested in the work of Robert Owen, the Utopian socialist whose reforms in his British factory towns showed promising results. Owen planned to transplant his work to the New World and he purchased the town of New Harmony, Indiana for his experiment. He collected a notable group of scientists, artists and educators in Pittsburgh and floated down the Ohio on a keelboat named the *Philanthropist,* but nicknamed "The Boatload of Knowledge". When the boat stopped in Steubenville on January 8, 1826, it left with Tappan's son Benjamin as an additional student. The experiment at New Harmony was not a success, but the younger Tappan apparently received a good education, as he later became a physician.

While Tappan would not name a town after himself, his family apparently had no problem with naming their children after each other as his father, eldest son and grandson all shared his first name. This was the only child he had by Nancy, who was plagued by poor health. She died in 1822, apparently of cancer. Tappan was devoted to her and mourned her loss, but in 1823 he remarried, to Betsy Frazer of Columbus. This union lasted until her death in 1840 and produced another son, Eli Todd Tappan. This son had brief careers as newspaper editor, lawyer and mayor of Steubenville before finding his true calling as an educator. He became a professor of mathematics, later served as President of Kenyon College, and was one of the first presidents of the National Education Association.

Tappan himself was not an imposing figure physically. Small in stature, his shock of hair turned white in later years, earning him the nickname "the hoary headed skeptic." He spoke through his nose with a sort of sing-song whiny voice that belied his considerable intellectual prowess. But it was Tappan's eyes that were his strongest physical trait. He was slightly cross eyed, with the left eye turning in. He made no effort to conceal this defect, and in fact used it to his advantage in court where he could appear to be looking at both judge and jury.

In his personal reading tastes, Tappan favored writers like Swift and Rabelais -- authors who shared his penchant for wit and sarcasm. Tappan had a generally low opinion of mankind and it was said that among his favorite words were "damnable" and "blockhead", such as when he called the legislature "a damnable set of blockheads." He was impatient and irascible, even with colleagues. Henry Howe uncovered one story, where when he was serving as judge, Tappan could not wait for another judge on the panel to return from a meal break. Upon hearing that the missing judge's saddlebags were under the bench, Tappan said, "I'll go on with my plea; they will do just as well."

These are not the traits of a successful politician. Tappan's contentious nature regularly doomed him to defeat at the hands of less able men. His friends advised him that to secure election he needed to "hermetically seal his vinegar cruet and cork up his damned sarcasms for a while," but he could not do it. In addition, he quarreled with the civic leaders who were his closest associates. He was successfully sued for slander by his own brother-in-law, and when James Wilson reported this in his paper, he broke with the editor as well.

Tappan ran for legislature, state senate, state Supreme Court, governor and U.S. Senator, and was defeated in all races, rarely even carrying his home county. Bezaleel Wells, a former partner turned bitter enemy, charged that Tappan "appears to possess an obliquity of intellect and perversity of disposition that keeps him constantly at variance with all the sober, reflecting and intelligent part of the community," adding that "he has never been able to get himself into office through the popular suffrages, although he has repeatedly attempted to force himself on the people."

The political parties in the country were in a state of realignment at this time, and Tappan came to align himself with the Jacksonian Democrats. The aristocratic Tappan was a most unlikely champion of the common man, but he came to associate the party conflict with the Federalist/Jeffersonian conflict of his youth, and he joined the party that better represented a break with his upbringing. Another colleague observed that Tappan "no doubt regrets that he cannot be with some party that is opposed to everybody, but as far as that is not possible he [inclines] to be for that among which he has no friends and can therefore safely call himself disinterested."

But Tappan's political fortunes began to rise after he started to support Andrew Jackson. He was a tireless behind-the-scenes worker with connections across the state and he began to amass political IOU's. A reward finally came on October 12, 1833, when President Jackson appointed him U.S. District Judge for the Ohio federal district. He assumed office immediately, but the job still required Senate confirmation in the spring. In the interim, his local opponents complained about a judge who openly said that the Scriptures were "all damned nonsense," and Southern senators were upset to hear that Tappan had spoken out in favor of slave rebellion. The Senate was feuding with Jackson on several other matters as well, and in May 1834 they soundly rejected Tappan's appointment by a 28-11 margin.

Tappan was now in his 60's and it looked as if his dismal political career was finished. But he had one last opportunity in 1838 when Ohio was to choose a U.S. Senator. At that time, senators were chosen by the state legislatures, and the Democrats were the majority party. The incumbent, Thomas Morris, was a Democrat, but he had alienated party officials with his strident abolitionist beliefs.

Abolitionism was the most controversial single issue of its day, and a candidate's position on it was used as a litmus test. Morris' extreme views meant that the legislature was looking for a moderate on the issue, and Tappan was able to pass as one by pulling his punches for once. Tappan was very much opposed to slavery but was not a part of the organized abolitionist movement. This was because his views were based on an Age of Reason natural law philosophy and he had little in common with the evangelical Christians that dominated the movement. So when asked if he approved of modern abolitionism, Tappan could honestly say no, and on December 20, 1838, he was elected to the U.S. Senate.

Ironically, two of the leading evangelical Christians in the abolitionist ranks were Tappan's brothers, Arthur and Lewis. Arthur was President of the American Anti-Slavery Society and the founder of Oberlin College, the first U.S. college to admit blacks. Lewis, the youngest of the Tappan brothers, played a major role in the *Amistad* affair, providing for financial and legal assistance for African slaves who had taken over their slave ship and landed in New York.

Despite their differences in age and religion, Benjamin and Lewis were close. Lewis constantly tried to convert his brother, and while Benjamin refused to be swayed, he at least spared Lewis his usual sarcasm. Of these attempts, Benjamin wrote to Lewis, "I do not dislike you for this, but I marvel that you do not tire in trying to proselyte me." When Benjamin wrote that he had been elected to the Senate, Lewis wrote back, "to tell you the honest truth I should have preferred to have heard you had become deacon of a church."

Even though he had attained office by soft peddling his stance on slavery, Tappan did not betray his principles. His most significant act as Senator occurred when he intentionally leaked the contents of a Senate treaty to the New York press. Secretary of State John C. Calhoun wanted to conduct negotiations for the annexation of Texas in secret, but Tappan was loathe to play such a role in adding another slave state to the Union. He arranged for Lewis to have the treaty printed, and for this he was censured by his Senatorial colleagues.

In less controversial matters, Tappan served on the Library Committee and worked on the Senate version of the bill that established the Smithsonian Institution. When he left for Washington, Tappan put his law practice in the hands of his partner, Edwin Stanton. He had known Stanton since the latter was a youth and was a classmate of Tappan's son, who later married Stanton's sister. Stanton loyally kept his mentor informed of local political developments, and Senator Tappan wrote back appreciatively, "I get more information from your letters as to Ohio matters than from all other sources."

When his term was up the Whigs controlled the legislature, so Tappan returned to Steubenville in 1845. In his last years he became a supporter of the anti-slavery Free Soil Party. He also devoted himself to his scientific pursuits and of course continued to freely dispense his irascible opinions. He died in 1857 without the deathbed conversion that Lewis had hoped for.

Benjamin Tappan was born three years before the Declaration of Independence and died three years before Lincoln's election. When he arrived in Ohio in 1799, his settlement was the only one for miles around, but when he died Ohio was a prosperous and progressive state on the verge of playing a national leadership role. He was the epitome of the kind of person that made these changes possible.

CHAPTER X
John Chapman: The True Story of Johnny Appleseed

One day in the late winter of 1805, five men came upstream on Duck Creek from Marietta. Most likely they traveled by poling a raft up the meandering creek that served as the main highway for new settlement in the area. This small group of pioneers was seeking to expand the frontier by settling farther upstream and their immediate goal was to find a site to build a new home. Near the present-day village of Dexter City, they chose a wooded hillside above Duck Creek as the site for their cabin. According to one account, they were seven miles from the nearest neighbor.

The five men were Nathaniel Chapman and four of his sons. Nathaniel was 58 and not in the best of health, but he was no stranger to the rigors of the outdoors. He had served as a Captain in Washington's army 28 years before, during the harsh winter at Valley Forge.

The oldest son from his second marriage, also named Nathaniel, was just as hardy. Though only 24, he had been living on the frontier and in the Marietta area for seven years. The other two sons from Captain Chapman's second marriage, Abner, 21, and Parley, who had just turned 20, were both able-bodied young men.

But it was John, the 30-year-old son from the Captain's first marriage, who had the most experience at frontier living. John Chapman had already spent several years on the edges of civilization and he would spend the next 40 years advancing those edges under the name of Johnny Appleseed.

The man who would become Johnny Appleseed was of average height and wiry in stature. He has been described as being about five-foot-eight inches tall and weighing approximately 125 pounds. His hair was long, fine and dark, and brushed behind the ears. At this point he was most likely clean-shaven, although he grew a beard in later years. One account mentions that he had hollow cheeks, a small mouth, and a small, turned-up nose. Nearly all accounts mention his dark, almost black, eyes that were both deep and sparkling. Reflecting the intensity that these eyes indicated, he was described as "quick and restless in his motions and conversation."

But what people noticed first about John Chapman was the way he dressed. He wore cast-off clothes of all kinds, from coffee sack shirts to several pair of ragged trousers pieced together. And while he usually wore an old wide-brim felt hat on his head, his feet were almost always bare, regardless of the weather or season. His garb has been described as "shabby and outlandish, even on the frontier."

This singularly dressed man and his father and half-brothers set about building a rude cabin on the site they had selected. The rest of the family waited for them in Marietta. This group included Captain Chapman's wife, Lucy, who was 42; her daughters Lucy, 17; Patty, 15;Persis, 11; Mary, nine; Sally, who was one year old; and younger sons Jonathan, seven, and Davis, four. The family was reunited as work progressed on the cabin, but John Chapman did not stay long. As soon as the new home was completed, he took a supply of apple seeds in leather pouches, put them in a canoe, and headed upstream on the Muskingum and Licking Rivers. While the rest of his family would stay in the area, Johnny Appleseed would follow the frontier until he became a legend.

There are few written records of Johnny Appleseed, and we cannot be absolutely certain it happened this way. The problem with legendary people is that they become larger than life and we can never be sure exactly what they really did do. Or, as the author Louis Bromfield observed, " the truth is, of course, that Johnny Appleseed has attained that legendary status where the facts are no longer of importance."

Schoolchildren today know of the kindly hermit Johnny Appleseed, who planted apple trees ahead of the pioneers to help pave the way for civilization. He is mentioned as part of our frontier folklore along with Paul Bunyan and John Henry. But, unlike these other two, Johnny Appleseed was a real person who spent most of his adult life in Ohio.

His legend grows because he left no written record, and the great storyteller impressed generations of children who embellished his stories with their own exaggerations until it appears he planted nearly every apple orchard in the Midwest and several more nationwide, and that he knew every major figure on the frontier from George Rogers Clark to Abraham Lincoln.

A look at the few direct observations available show that most popular stories about Johnny Appleseed have no known basis of truth. But a look at what we do know shows a man of intriguing contrasts and depth.

Although he was thought by his contemporaries to be eccentric and even shiftless, he had a burning sense of purpose in his dual goals of providing the pioneers with apple trees and spreading the gospel according to the Swedish theologian and mystic Emmanuel Swedenborg. He was a zealous missionary for both causes, but he was also a businessman grounded in the real world who owned, at one time or another, 22 separate tracts of land totaling nearly 1200 acres. Although he was a wandering, dreamy storyteller to the children he loved so, he was also a conservationist and herbalist who lectured adults on the merits of his lifestyle of simplicity and self-reliance.

He left almost nothing in writing, yet he was well-read and universally esteemed for his intelligence, despite a lack of formal education. And while he introduced apples to much of the frontier, he also introduced the notorious pest dog fennel in a misguided attempt to provide an herbal cure for malaria. He was a pacifist who opposed the unnecessary killing of any living creature, yet he was a war hero. He was gentle and kind and was compared to St. Francis of Assissi, yet he could be humorous and earthy as well. And while he was almost nothing like any of his fellow frontiersmen, he was loved and respected by all of them.

John Chapman was born on September 26, 1774, in Leominster, Massachusetts. He was the first son and second child born to Nathaniel and Elizabeth Simonds Chapman. An older sister

Elizabeth was born in 1770; she was to be the only member of John's family not to move west.

The Chapman family had been in America since 1639. The Simonds family had been in Massachusetts since 1635 and was apparently better educated and more well-to-do than the Chapmans. A first cousin of Elizabeth Simonds was Benjamin Thompson, who would gain some renown as a scientist and nobleman. He could have been as well-known in America as Benjamin Franklin except he made the unfortunate choice of supporting the British during the American Revolution. He fled the country after the war, eventually married a countess and took the title Count Rumford.

John Chapman was born on the eve of the Revolution and his father was a member of the local company of Minute Men. When the British marched on Lexington and Concord in April of 1775, Nathaniel's unit was called out, although they arrived too late to participate in the fighting. However, Nathaniel stayed and became part of Washington's Army. He fought at Bunker Hill and eventually rose to the rank of captain. He remained in the army until 1780, so John hardly saw his father from the time he was seven months old until he was six years old.

John Chapman's mother died not long after her husband left to go to war. She was pregnant and in ill health in June of 1776 when she sent this letter to her husband, who was fighting in the New York campaign:

Loving Husband,

These lines come with my affectionate regards to you hoping they will find you in health, tho I still continue in a very weak and low condition. I am no better than I was when you left me but rather worse, and I should be very glad if you could come and see me for I want to see you.

Our children are both well thro the Divine goodness.

I have received but 2 letters from you since you went away-neither knew where you was till last Friday I had one and Sabathday evening after, another, and I rejoice to hear that you are well and I pray you may thus continue and in God's due time be returned in safety.

I have wrote that I should be glad you could come to see me if you could, but if you cannot, I desire you should make yourself as easy as possible for I am under the care of a kind Providence who is able to do more for me than I can ask or think and I desire humbly to submit to His Holy Will with patience and resignation, patiently to bear what he shall see fitt to lay upon me. My cough is something abated, but I think I grow weaker. I desire your prayers for me that may be prepared for the will of God that I may so improve my remainder of life that I may answer the great end for which I was made, that I might glorify God here and finally come to the enjoyment of Him in a world of glory, thro the merits of Jesus Christ.

Remember, I beseech you , that you are a mortall and that you must submit to death sooner or later and consider that we are always in danger of our spiritual enemy. Be, therefore, on your guard continually; and live in a daily preparation for death- and so I must bid you farewell and if it should be so ordered that I should not see you again, I hope we shall both be as happy as to spend an eternity of happiness together in the coming world which is my desire and prayer.

So I conclude by subscribing myself, your ever loving and affectionate wife,
Elizabeth Chapman.

These brave words were written on June 3, 1776. On June 26, Elizabeth gave birth to a son, who died shortly afterwards. On July 4, the Declaration of Independence was signed, and on July 18, Elizabeth Chapman died at the age of 28, leaving a 5-year-old daughter and a 21-month-old son.

Elizabeth's parents took over the care of young Johnny and his sister. Nathaniel remained in the army until 1780, when he was discharged for mismanagement of military stores. It's not certain whether it was his competence or integrity that was questioned, but because of the nature of his discharge he was ineligible for pension and land bounties that were made available to other veterans.

In that same year, Nathaniel married 18-year-old Lucy Cooley of Longmeadow, Massachusetts.

They settled in that town and had ten children in the next 22 years. It was here that Johnny spent his formative years and got the only formal education he was ever to have. According to some accounts John was apprenticed to an orchardist here, but the truth is there is no written record of John Chapman from his birth until 1796.

In that year, John turns up in Warren, Pennsylvania, with his half-brother Nathaniel. One can only assume that wanderlust and an overcrowded home caused John to move west. This region on the Allegheny River was very unsettled at the time and John was among the first to arrive. He owned no property but was a squatter in an area where land speculation and claim jumping were rampant. There were also many Indians still living in around here and John may have learned wilderness survival skills from them.

Warren and Franklin, Pennsylvania, were apparently the center of John's activities from 1796 to 1804. But he probably first visited Ohio at the beginning of this period. Nathaniel moved to the Waterford settlement near Marietta in 1797, so John was certainly familiar with this area.

There is some evidence to indicate he had already begun his orcharding career during this period. He is on record as dealing in apple trees at the trading post in Franklin, and his first orchard in Ohio is generally accepted as being planted near Steubenville in 1801. His method of planting was to get seeds from cider presses and take them downstream in a canoe. This work gradually took him further west into more unsettled areas.

It is also likely that during this period John became involved with the other passion in his life--the Swedenborgian religion. It was probably Judge John Young of Greensburgh, Pennsylvania, who introduced John to the writings of Emanuel Swedenborg (1688-1772). Swedenborg was a Swedish scientist and nobleman who turned from science to devote the last 28 years of his life to voluminous interpretation of the Bible.

According to Swedenborgian theology, there are simultaneous physical and spiritual worlds, but the spiritual world goes on after physical life has ended. Christ's humanity rather than divinity is stressed, as adherents believe that Jesus became divine by living the perfect life. The goals of the religion are attainment of rational knowledge and of a perfectly loving spirit. Swedenborgianism has elements of mysticism, but also was a much more rational and refined religion than was usually found on the frontier. While the church never attracted a large membership, it did contain members who were said to be of excellent intellect.

Johnny Appleseed was apparently one of these. While he never attended Harvard, as some legends claim, and there is very little record of any writings and correspondence by him, he nonetheless was considered quite intelligent by several observers who did take notes. Frontier religion tended to be emotional and superstitious, and the populace was probably more receptive to Johnny's apples than his erudite religion. But in a time and place where literacy was rare, and preaching and oratory were often all that could pass as culture, it is a measure of the respect accorded Johnny Appleseed that he was asked to give the Fourth of July oration near Norwalk in 1816.

There are other characteristics of the frontier that need to be understood in order to put his work in the proper context. One of these is that no one had very much hard cash, and in many cases a primitive barter economy was a major way of doing business. The tremendous amount of land opening for settlement in the West meant little to would-be pioneers who had no cash to put down. Title to much of the new lands was held by eastern speculators, and land was often available only in large parcels that few commoners could afford.

Such a system encouraged claim jumpers and "squatters", who simply built a home on a property without regard for who held title to it. Litigation might take years and by then the squatters may well have moved farther west. Revolutionary War veterans were often given generous land bounties, but others had to take their chances as squatters or else come up with some cash.

Another salient characteristic of the frontier was the incredibly poor transportation system. Roads were rough where they existed at all and passage across the Appalachian Mountains was difficult and expensive for people and even more so for property.

Rivers and streams were the frontier version of freeways. Streams cut through the hills at their weakest points and valleys alongside offered the best farmland, so it is not surprising that new settlements were located there. So much easier and cheaper was water travel than overland routes that settlers preferred to float their produce 2000 miles down the Ohio and Mississippi Rivers rather than risk crossing the mountains.

This long trip meant that farmers seeking to sell their crops had to put their produce in a form that would survive such a trip without spoilage. This was where the ability to make "moonshine" became a valuable skill, as corn turned into whiskey was a form of liquid asset that could make this journey.

Apples were another crop that held up well in various forms. They could be used for apple cider or apple butter, or dried and preserved and used throughout the year. Apple trees survived all sorts of conditions, and seeds were plentiful. Apple trees were considered to be a sign of civilization. The Ohio Company had a stipulation that to lay claim to 100 acres, one needed to put out 50 apple trees and 20 peach trees within three years.

In 1864, the Reverend Henry Ward Beecher, a renowned clergyman today best remembered as the father of the author of *Uncle Tom's Cabin,* delivered a speech entitled "The Political Economy of the Apple." Beecher praised the apple as "beyond question, the American fruit... whether neglected, abused, or abandoned, it is able to take care of itself and to be fruitful of excellences. That is what I call being democratic." In discussing "the apple as an article of commerce," Beecher stated that "the apple comes nearer to universal uses than any other fruit of the world."

The one problem associated with apple trees on the frontier was that it took a few crucial years before they would bear fruit. The self-appointed mission of Johnny Appleseed became to have those apple trees ready for the first settlers. Other orchardists worked in early Ohio, but only Johnny Appleseed moved with -- or rather, just ahead of -- the frontier.

Between the years of 1797 and about 1804, Johnny maintained a dual residency in western Pennsylvania and Ohio. Gradually he moved his operations into central Ohio. Some of his early known orchards in Ohio were near Morristown in Belmont County, in Carroll County, and in Licking County. Eventually he went up the Muskingum River to Coshocton and then up the Walhonding and Kokosing Rivers to Mount. Vernon in Knox County.

He was selling apple trees in Mount Vernon in 1806 and voted there that year. In 1809, he bought two town lots in Mount Vernon. But the plan of Johnny Appleseed was not to build a home and settle down. These lots were purchased to house permanent orchards to grow stock in.

He also kept many orchards in other places as well and came to expand his work to the Mohican River and Black Fork in the Mansfield and Ashland area, where he remained a regular sight for several years. He was living in this area when the War of 1812 broke out.

Up until this time, Indians and settlers had lived side by side in uneasy truce since the Greenville Treaty of 1795 had ended Indian wars in Ohio. But when war with Great Britain was declared, many Indians were incited to side with the British and the threat of renewed hostilities loomed large. Most Indians who lived in the Mansfield area left for their own protection and at least one friendly Indian who stayed was murdered.

Tension was high in the area on August 21, when word came that Detroit had fallen to the British and that nine ships were sailing from Detroit to Cleveland on Lake Erie for an invasion. Johnny Appleseed spread the alarm to settlers in the area to go to Mansfield for safety. It turned out the nine ships were carrying prisoners of war and no invasion was imminent, but that made the fears expressed no less real.

Then, in early September, a trader from Mansfield was killed and scalped by Indians. Settlers again gathered in town and Johnny Appleseed was again selected to spread the alarm. He covered the 30 miles from Mansfield to Mount Vernon, warning all along the way. The next day he returned with soldiers from Mount Vernon.

Traditional versions of the story have Johnny Appleseed running thirty miles barefoot in the night, calling out "The spirit of the Lord is upon me, and he hath anointed me to blow the trumpet in the wilderness, and sound an alarm in the forest; for, behold, the tribes of the heathen are round about your doors, and a devouring flame followeth after them!" Since speed was of the essence in warning of Indian attacks, it is more likely he rode a horse and shouted a somewhat briefer alarm. Regardless of the particulars of the incident, the citizens' choice of Johnny Appleseed as messenger establishes him as a war hero.

There was no full-scale attack that ensued, but the following week a minister named Copus, who was friendly to Indians, felt it was safe to return to his farm. His group was attacked by a band of Indians and Copus and a few others were killed.

After the War of 1812, Johnny Appleseed began to buy more property for his orchards until he owned 640 acres by 1815. His usual method of operation was to clear the land and build a rude dwelling to suit homesteading rules and meet the neighbors, giving portions of Swedenborgian tracts to the adults and telling stories to children. He would then plant his orchards and move on. There seems to be some question whether he believed in grafting apple trees. Several accounts mention that he was a purist who felt grafting was unnatural. But Swedenborg, in his writings, mentions grafting as a positive metaphor several times, and since the technology to do so was both available and logical, it is likely that Johnny practiced it.

Gliding to a Better Place

Sometime in the 1820's, Johnny began branching westward from the Mansfield area. He turned up in places like Urbana, Mount Blanchard, and Van Wert in western Ohio, and was seen in Fort Wayne, Indiana, as early as 1830. The northwest region of Ohio, called the Black Swamp, was the last area of the state to be settled. Johnny's routes in this area once again followed the rivers, as he world go downstream on the Blanchard and Auglaize rivers to the Maumee River at Defiance, then upstream on the Maumee to Fort Wayne. As he had done previously, he had a sort of dual residency in both the Mansfield and Fort Wayne areas from about 1827 to 1840. He would make regular cyclical visits to his orchards in both places and points in between. In the 1830's he also bought several acres of land in Indiana and gradually came to live there year-round.

As he grew older, Johnny's irregular comings and goings helped him grow to living legend status. His name even changed to reflect this. As a younger man, John Chapman had been referred to as Appleseed John. This gradually became John Appleseed. It was not until his later years, as a bearded and eccentric wanderer, that he was called Johnny Appleseed. After his death, he was exclusively referred to as Johnny Appleseed by those who remembered him as an old man, and these people spread his legend farther than the man had ever been in his lifetime.

It is in legend that Johnny Appleseed lives on best. Tales and legends that illustrate his character are as abundant as actual records are sparse. Many of them stress his gentleness and respect for all life.

Not only was Johnny opposed to hunting, he hated killing any living creature. Stories are told about the great guilt he suffered when in anger he killed a snake that had bit him. He once removed a yellow jacket from his shirt after it had stung him, but refused to kill it saying it was only obeying its nature and meant him no harm. He even put out his campfire one night because mosquitoes were flying into it, saying, "God forbid that I should build a fire for my own comfort that should be the cause of destroying any of His other created works."

No doubt his religious beliefs were largely responsible for such high standards of behavior. He was always eager to distribute his religious literature which he called "Good news fresh from Heaven." He even wrote to the head of the Swedenborgian Society in Philadelphia offering to exchange land he owned for more literature. The books that he did have he often distributed by tearing them into sections and leaving them at places he visited.

While he was not a teetotaler, he never drank to excess. The use of coffee, tea and tobacco he shunned entirely, having a preference for milk and honey. He was not a complete vegetarian, but ate lots of nuts and berries, and on one occasion disappointed a worker he'd hired by offering him a "meal" of black walnuts. He was a firm believer in the medicinal value of herbs and, in addition to apple seeds in his leather pouches, he also carried such herbs as horehound, catnip, pennyroyal, ginseng, goldenseal, wood bitney and dog fennel. He lived simply and self-reliantly and deplored waste of any kind. When he stayed anywhere he usually insisted on sleeping outside, and at mealtimes would not eat unless all children present were served first.

He was especially fond of children, particularly the girls. He would gather children around him and regale them with stories, like the one about the time he was canoeing on an icy river and decided the ice floes moved faster than his craft. He put his canoe on one and promptly fell asleep only to find himself miles past his destination when he awoke. These young children would tell such tales to their grandchildren in the years to come and the Appleseed legend would grow.

Despite the popular notion that prevails, there is no evidence that Johnny Appleseed ever wore a mush pot for a hat. But he did go barefoot most of the time, and his attire of cast-off clothes and his unusual habits certainly marked him as a frontier eccentric. His behavior was so strange that some sought to explain it by repeating the story that as a young man he had been kicked in the head by a horse.

When asked why he had never taken a wife, he replied that he would have a mate in Heaven. He believed that whatever we did in the physical world, we would continue to do in the spiritual world. This caused one wag to remark about himself that he would surely be unemployed in the afterlife because he was a grave digger. But this was a good-natured jibe, for Johnny had the respect and admiration of his peers, despite his unusual ways.

Besides, he could give as good as he got. A favorite story concerns a long-winded preacher who gave a sermon in Mansfield. In some versions, the preacher was famous circuit rider Peter Cartwright. This minister was loudly rebuking the citizens of Mansfield for their decadent ways, saying how far they had strayed from Biblical ideals. At one point far into his sermon he called out, "Where now is there a man who, like the primitive Christians, is traveling to heaven clad in coarse raiment?" At this, Johnny Appleseed walked towards the man, looking for all the world like John the Baptist, and replied, "Here is your primitive Christian." The flustered preacher soon lost his thread of concentration and the sermon ended shortly thereafter, making Johnny a hero once again.

There are many more stories about Johnny Appleseed, and he would have hardly had the time to do all that was said of him. It is possible that in his later years he may have traveled as far west as Iowa once, but he couldn't possibly have been to all the places and met all the people that later legends have him doing. Since many people only saw him at a certain time of the year, there may have been a lot of open speculation about where he had been in the interim. Of course, where he had been was tending orchards elsewhere, inspiring legends to a different set of people.

Johnny Appleseed probably visited his family back on Duck Creek on a semi-regular basis, as it was on his likely seed route for a while. His last known visit to southeastern Ohio was in 1842. Johnny's brother-in-law John Whitney had a cousin named W. M. Glines who was present when Johnny Appleseed last visited his family. Thirty-three years later he wrote an account of that meeting that serves as an insightful portrait of Johnny as a 68-year-old man, approaching the end of his mission.

Glines writes of Johnny Appleseed:

His last visit to Ohio was in October, 1842; Mr. Nathaniel Chapman was a neighbor to me at the time and a very warm friend with all; Johnny made it his home with him while on his visit to his friends. A Mr. John Whitney who married Sally, lived in the neighborhood, was a farmer, and during the summer the lightning had struck a very large black oak tree on Whitney's land and knocked it to pieces from the top to the roots; and some of the fragments were converted into some very comfortable sized rails, and of such length as made them convenient for that purpose.

Johnny having heard of the circumstance and that Whitney had laid them up in his fence, he and Nathaniel came to my house and would have me go with them to see the rails that were made by lightning; when we got ready to start I proposed taking my gun along to kill some squirrels or rabbits; to this Johnny demurred; he read me a severe lecture upon the subject of taking life from any living creature; he maintained that God was the Author of all life, hence it belonged to Him whenever He was ready to demand it, and in as much as we could not give life to any creature, we were not at liberty to destroy life with impunity; after his lecture and to please him, I put up the gun and we moved on.

...Until we arrived at the fence in question.
He commenced an examination, he measured them, counted them, and viewed the roots from whence they came; he then turned to Mr. Whitney and read a sermon upon the wonderful Providence of God to man, Said he, God has given you a large family of boys, they have cleared you a large farm in the woods, and they have worked hard to do it. Making rails is hard and heavy work. God pitied the boys, hence He sent lightning to make your rails and He selected that hard old burley tree close by where you most needed them, now said the old man, can't you see it?"

Whitney hung his head for a moment, then replied, that he always tried to be thankful to God for His kind care over him and his family, but that he never heard of His making rails for anybody before.

In addition to this revealing anecdote, Glines also offers observations on Johnny's attire and views on it:

He was very much disturbed with the extravagance of the world at large, particularly in the article of dress. He would often have one leg of pantaloons made of one kind of color and cloth and the other entirely of another kind and quality. Buckskin was his favorite foot covering. Shoes he abhorred. While on his last visit his niece, Miss Rebecka Chapman (a daughter of Nathaniel's) made him a shirt, one half calico, the other muslin. On the one of the muslin, were two large letters, perhaps A.D. These he had so arranged that one was on either side of the bosom. That seemed to please him.

After this visit, Johnny returned to Fort Wayne where he spent the remainder of his life. But he did not return there to retire. Johnny Appleseed continued to work in his orchards as he seemingly always had done. Just over two years after his last visit to his family along Duck Creek, Johnny Appleseed became ill while out working in late winter. He went to the cabin of a family named Worth, where he died on March 18, 1845, at the age of 70.

There was a flattering but brief obituary in the Fort Wayne newspaper but otherwise his death went unnoticed. It did, however, take several years to straighten out the estate, as Johnny had owned over 400 acres in various locations at his death. But gradually it became apparent that Johnny had been right all along: his spirit did live on after his physical life had ended.

An article in *Harper's Monthly Magazine* in 1871 brought national attention to Johnny

Appleseed, but it was the memories of children he'd entertained that really kept his spirit living. Stories have been told through several generations about Johnny Appleseed, and a series of literature has been written about him. There were some romantic novels around 1900 that stressed his career choice arising from a frustration in romance. Poets from Vachel Lindsay to numerous local authors have sung his praises in verse. Many children's books add to the folklore available. Pulitzer Prize winning author Louis Bromfield, a native of Mansfield, called Johnny Appleseed "an ideal historical point of reference for rural Ohioans."

Today, Johnny Appleseed's spirit lives on through this literature, through folklore handed down, and through monuments erected to him in Fort Wayne, Mansfield, Ashland, Leominster and Dexter City. And he will continue to live on as long as the descendants of his apple trees continue to provide us with the fruits of his labors.

SELECTED BIBLIOGRAPHY

Ambler, Charles H. *George Washington and the West.* New York: Russell & Russell, 1971 repr. after 1936 by University of North Carolina Press.

Baldwin, Leland D. "Shipbuilding on the Western Waters." *Missisipi Valley Historical Review*, vol. 20 (June, 1931) : 29-44.

Banta, R. E. *The Ohio.* New York: Rinehart & Co., 1949.

Boatner, Mark Mayo III, *Encyclopedia of the American Revolution.* New York: David McKay Co. June, 1966.

Bond, Beverly, Jr. *The Foundation of Ohio.* Columbus: Ohio Historical Society, 1944.

Booth, Russell, Jr. *The Tuscarawas Valley in Indian Days.* Cambridge, Oh: Gomber House Press, 1994.

Brown, Jeffrey P. and Andrew R. L. Cayton. *The Pursuit of Public Power: Political Culture in Ohio 1787-1861.* Kent, Oh: Kent State University Press, 1994.

Butterfield, Consul Willshire, *History of the Girtys.* Cincinnati: Robert Clarke & Co., 1890.

Cammarata, Kathy. "Won't You Come Home, Anne Bailey?": *Southeast Ohio*, vol. 29, no. 3 (Summer/Fall 1997): 35-37.

Chandler, David Leon. *The Jefferson Conspiracies: a President's Role in the Assassination of Meriwether Lewis*. New York: William Morrow & Co, 1994.

Cleland, Hugh, *George Washington in the Ohio Valley*. Pittsburgh: University of Pittsburgh Press, 1955.

Crain, Ray, *The Land Beyond the Mountains*. Urbana, Oh: Main Graphics & Associates, 1994.

Dexter, Ralph W. "Benjamin Tappan, Jr. As a Naturalist and a Malacologist". *Storkiana*, vol. 41 (March 1971): 45-49.

Dowd, Gregory E. *A Spirited Resistance: The North American Indian Struggle for Unity 1745-1815*. Baltimore: Johns Hopkins Press, 1992.

Downes, Randolph C. , *Council Fires on the Upper Ohio*. Pittsburgh: University of Pittsburgh Press, 1940.

Downes, Randolph C. , *Frontier Ohio 1788-1803*. Columbus: Ohio State Archaeological & HistoricalSociety: 1935.

Gibson, John B. "General John Gibson." *Western Pennsylvania Historical Magazine.* (October, 1922): 298-310.

Gilbert, Bil. *God Gave us This Country: Tekamthi and the First American Civil War.* New York: Macmillan, 1989.

Hanko, Charles W. *The Life of John Gibson.* Daytona Beach, Fla: College Publishing Co., 1955.

Hatcher, Harlan. *The Western Reserve: The Story of New Connecticut in Ohio.* Kent, Oh: Kent State University Press, 1991.

Heckewelder, John. *History, Manners and Customs of the Indian Nations who once inhabited Pennsylvania and the neighboring states.* Philadelphia: Abraham Small, 1819.

Hildreth, Samuel P., M.D. *Memoirs of the Early Pioneer Settlers of Ohio.* Cincinnati: Clearfield Co., 1854.

Hooper, Osman Castle. *History of Ohio Journalism.* Columbus: Spahr & Glenn Company, 1933.

Howe, Henry. *Historical Collections of Ohio.* Cincinnati: Krehbiel & Co., 1888.

Hulbert, A.B. "Western Ship-Building." *American Historical Review* vol. 21 (October, 1895): 720-733.

Kellogg, Louise Phelps. *Frontier Advance on the Upper Ohio.* Madison, Wis.: State Historical Society of Wisconsin, 1916.

Kelsay, Isabel Thompson. *Joseph Brant: A Man of Two Worlds.* Syracuse, N.Y.: Syracuse University Press, 1984.

Kohn, Richard, H. "General Wilkinson's Vendetta With General Wayne: Politics and Command in the American Army 1791-1796. *The Filson Club History Quarterly.* (October, 1971): 361-372.

Lewis, Virgil Anson. *Life and Times of Anne Bailey.* Charleston, WV: Butler Printing Co., 1891.

Northwest Territory Celebration Commission, *History of the Ordinance of 1787 and the Old Northwest Territory.* Marietta, Oh: Northwest Terrritory Celebration Commission, 1937.

Olmstead, Earl P. *David Zeisberger: A Life Among the Indians.* Kent, Oh: Kent State University Press, 1997.

Parkman, Francis. *The Conspiracy of Pontiac.* NY: Collier Books, 1967.

Pieper, Thomas I. & James B. Gidney. *Fort Laurens 1778-1779: The Revolutionary War in Ohio.* Kent, Oh: Kent State University Press, 1976.

Price, Robert. *Johnny Appleseed: Man and Myth.* Bloomington, Ind.: Indiana University Press, 1954.

Quaife, M. M. ed, "General James Wilkinson's Narrative of the Fallen Timbers Campaign." *Mississippi Valley Historical Review.* : 81-90.

Ratcliffe, Donald J., ed. "The Autobiography of Benjamin Tappan." *Ohio History*, vol. 85 (1976): 109-157.

Shreve, Royal Ornan. *The Finished Scoundrel.* Indianapolis: Bobbs-Merrill, 1933.

Sibley, William G. *The French 500.* Gallipolis, Oh: Gallia County Historical Society, 1933.

Smith, Thomas H. Ph.D, ed. *Ohio in the American Revolution.* Columbus, Oh: Ohio Historical Society, 1976.

Stille, Samuel H. *Ohio Builds a Nation.* Lower Salem, Oh: Arlendale Book House, 1953.

Sugden, John. *Tecumseh: A Life.* New York: Henry Holt, 1998.

Thwaites, Reuben Gold and Louise Phelps Kellogg, *Documentary History of Dunmore's War.* Madison, Wisc.: Wisconsin Historical Society, 1905.

Van Every, Dale, *A Company of Heroes: The American Frontier 1775-1783.* New York: William Morrow & Co., 1962.

Van Every, Dale, *Forth to the Wilderness: The First American Frontier 1754-1774.* New York: William Morrow & Co., 1961.

Van Trees, Robert V., *Banks of the Wabash.* Fairborn, Oh: Apollo Books, 1986.

Wallace, Paul A. W., *Thirty Thousand Miles with John Heckewelder.* Pittsburgh: University of Pittsburgh Press, 1958.

Wiley, Richard T. "Ship and Brig building on the Ohio and its Tributaries." *Ohio Archaeological and Historical Quarterly*, vol. 22: 54-64.

Wyatt-Brown, Bertram. *Lewis Tappan and the Evangelical War Against Slavery.* Cleveland: Case Western Reserve University Press, 1969.

INDEX

Adams, John, 166
Adena, 151
Albany, NY; 91, 130, 133
Algonquin tribe, 119
Age of Reason, 211
Allegheny Cemetery, 47
Allegheny County, PA, 40
Allegheny Mountains, 55, 76, 171
Allegheny River, 222
American Bible Society, 117
American Philosophical Society, 71
Amistad, 212
Appalachian Mountains, 169, 224
Appleseed, Johnny: see John Chapman
Arnold, Benedict, 90
Articles of Confederation, 12
Ashland, 226, 236
Auglaize River, 68, 106, 119, 138, 229
Augusta County, VA, 75
Aurora, 206

Bailey, Anne Hennis Trotter, 73-88
Bailey, John 76, 85
Bainbridge, 145
Baker's Bottom Massacre, 23
Baltimore, MD 189
Barlow, Joel, 78
Beaver River, Pa 54-55
Bedford, England, 49

Beecher, Henry Ward, 225-226
Bellefontaine, 143
Belmont County, 226
Belpre, 111
Bethlehem, PA 51, 52, 53, 61, 65, 70, 71
Biddle, Charles, 110
Big Bottom Massacre, 99
Black Hoof, 149
Black Swamp, 229
Blackfish, 143
Blanchard River, 229
Black Fork, 226
Blenheim, Battle of, 74
Blennerhassett, Harmon, 111, 113-14, 175
Blennerhassett, Margaret, 111, 175
Blennerhassett Island, 111, 113, 175
Bloody Bridge, Battle of, 126
Blue Jacket, 145, 150
Blue Licks, 143
"Boatload of Knowledge" 208
Bolivar, 51
Boone, Daniel, 143-144
Boonesborough, 143
Boston, MA, 13, 167, 182, 188, 193
Boston, 201
Boston Post Boy and Advertiser, 162
Boston Tea Party, 163, 188
Boswell, James, 132
Bouquet, Henry, 10, 23, 127
Braddock, Gen. Edwin, 8, 75

Braddock's Defeat, 10, 120
Bradstreet, Colonel John, 127
Brant, Isaac, 136-37, 140
Brant, Joseph, 37, 119, 128-140;
Brant, Molly, 130
Brantford, Ontario, 140
Brock, Isaac, 155-156
Brodhead, Colonel Daniel, 35-38
Bromfield, Louis, 217, 236
Buck Creek, 146
Bunker Hill, Battle of; 164, 225
Burgoyne, General John, 91, 133
Burr, Aaron, 90, 109-14, 175
Butler, Simon: see Simon Kenton
Butler, Walter, 133, 137

Cahokia, IL; 30, 128
Calhoun, John C., 212
Calvinism, 197
Camp Charlotte, 25, 28
Campus Martius, 184
Canada, 49, 90, 109, 116, 120, 132, 156
Captives' Town, 63
Caribbean Sea, 162
Carpenter, Joseph, 184, 187
Carroll County, 226
Cartwright, Peter, 232
Cave-In Rock, 176
Celeron de Blaineville, 4
Centinel of the Northwest Territory, 105, 183
Champlain, Samuel De, 119

Chapman, Abner, 215
Chapman, Davis, 216
Chapman, Elizabeth, 219
Chapman, Elizabeth Simonds, 218, 219
Chapman, John, 85, 214-236
Chapman, Jonathan, 216
Chapman, Lucy, 216
Chapman, Lucy Cooley, 216, 221
Chapman, Mary, 216
Chapman, Nathaniel, Jr., 222, 233
Chapman, Nathaniel, Sr., 215, 221
Chapman, Parley, 215
Chapman, Patty, 216
Chapman, Persis, 216
Charleston, SC; 167
Charleston, WV; 74, 76-77, 81, 86
Cherokee tribe, 141
Cherry Valley Massacre, 133
Cheesekaw, 142, 145
Chester, PA; 168
Chicago, IL; 46, 156
Chillicothe, 25, 111, 142-43, 150, 184, 187-93, 207
Chippewa tribe, 120
Cincinnati, 66, 98, 173, 175, 183-87, 189, 191-92
Cincinnatus, 178
Clark County, 146
Clark, George Rogers, 25,30, 35-37, 96-97, 102, 143, 157, 217
Clark, William, 73, 111, 173

Gliding to a Better Place   **245**

Cleaveland, Moses, 200, 203
Cleveland, 120, 201, 205, 227
Cole, Thomas, 196, 207
Clendenin, Colonel, 77, 81
Clinton, Governor of NY, 138
Columbus, 165
Columbus, 208
Committee of Correspondence, 28
Comstock, Dr.; 169
Concord, Battle of, 164, 225
Connecticut, 192, 195, 198, 200
Connolly, Dr. John, 24, 28, 29
Constitution, U. S.; 14, 15, 94, 97, 114
Constitutional Convention, 14
Continental Congress, 12, 13, 29, 33, 59, 90, 92, 162, 165
Conway, General Thomas, 92
Cooper, Mary Ann, 85
Copus, Reverend, 228
Cornstalk, 24, 25, 29, 31, 45, 143
Cornwallis, surrender, 22, 62
Corydon, 47
Coshocton, 33, 37, 58, 60-1, 127, 226
Craik, Dr. Robert, 9
Cranston, RI; 167-68
Crawford, William, 9, 11, 39-40,64
Creek tribe, 153
Cresap, Colonel, p. 26-27
Cramer, Zadock, 178
Croghan, George, 127, 157
Cumberland, MD; 7

Custer's Last Stand, 100
Custis, Martha, 9
Cutler, Manasseh, 13, 15
Cuyahoga River, 65, 120-21, 129, 135, 200, 201, 202

Dalyell, Captain James, 126
Dartmouth College, 128, 131
Dartmouth, Lord, 164
Dayton, Jonathan, 110, 114
Dayton, 110, 186
Declaration of Independence, 27, 162, 164, 213, 221
Defiance, 229
Delaware tribe, 23, 29-38, 51-62, 72, 149
Delaware Prophet, 122
Detroit, MI; 29-38, 46, 56-58, 61-68, 102, 108, 121-27, 137,151,15657,170,205,227
Detroit River, 122, 126, 134
Devol, Jonathon, 174
Dexter City, 215, 236
Dish with One Spoon Concept, 151
Duck Creek, 214-15, 233, 235
William Duer, 78
Duncan Falls, 136
Dunlop's Station, 99
Dunmore, Lord, 25, 28, 29
Dunmore's War, 19, 24, 26-7, 30, 75, 132

East Liverpool, 9
Eaton, 103
Ecorces River, MI, 123
Edwards, Jonathan, 197

Elizabeth, PA; 172
Elliot, Matthew, 29, 58, 61
Embargo Act of 1808, 178
Enabling Act, 192
England, 4, 7, 9, 31, 51, 95, 116, 120, 130, 135, 155
Erie, Pa; 125, 157, 170, 202
Erie Triangle, 41

Fairfax, Lord, 3,5
Fairfield, Ontario, 66, 68, 69, 71
Fallen Timbers, Battle of, 16, 68, 107-108, 139, 146, 170
Falls of the Muskingum, 136
Falls of the Ohio at Louisville, 111, 175, 176
Firelands, 199
Florida, 109, 114
Fly, 2
Forbes, General John, 8, 9, 21
Fort Chartres, 126
Fort Dearborn, 156
Fort Defiance, 146
Fort Duquesne, 8, 21
Fort Finney, 144
Fort Gower, 27
Fort Gower Resolves, 27
Fort Greeneville, 105-6, 149
Fort Hamilton, 99, 103
Fort Harmar, 41-2, 67, 84, 136
Fort Harrison, 47
Fort Henry, 186
Fort Jefferson, 44, 66, 99, 103
Fort Laurens, 18, 33-5, 47, 60
Fort LeBoeuf, 5, 10
Fort Lee, 76, 77, 81, 82

Fort Massac, 111
Fort Malden, 151, 155, 158
Fort Meigs, 156
Fort Miami, 105-106, 139
Fort Miamis, 125
Fort Michilimackinac, 46, 123, 125
Fort Necessity, 7,9
Fort Niagara, 126, 131, 133
Fort Ontario, 127
Fort Pitt, 8, 10, 19-24, 28-42, 53, 56-60, 102, 123, 125, 127
Fort Randolph, 31, 78
Fort Recovery, 105, 146
Fort Sandusky, 124
Fort Savannah, 76
Fort Stanwix, 133, 142
Fort St. Clair, 103
Fort Stephenson, 157
Fort Ticonderoga, 91, 131
Fort Venango, 5, 125
Fort Washington, 98, 99, 101, 103
Fort Wayne, 98, 137,146,149, 152, 155,146, 229, 235, 236
France, 3, 4, 9, 91, 109
Franklin, Benjamin, 8, 165-6, 196, 197
Franklin, PA; 222
Freeman, Edmund, 186-188
Freeman's Journal, 186
French Revolution, 111

Gallatin, Albert, 44
Gallipolis, 74, 78, 79, 85-87
Galloway, Albert, 146
Galloway, Rebecca, 147

Gliding to a Better Place

Gamecock, 162
Gaspee, 163
Gates, General Horatio, 90-93
General Washington, 158
George III, 54, 128, 132
Georgia, 52, 141
Ghost Shirt, 148
Gibson, George, 21, 42
Gibson, John, 18-48, 60, 66, 152, 192
Gilman, Benjamin Ives, 173
Girty, George, 29, 37
Girty, James, 29
Girty, Simon, 18, 20, 25, 27, 29, 31-35, 38, 40, 42-3, 46-7, 49, 58, 66, 134, 138, 143
The Glaize, 68, 138-9
Gladwin, Major Henry, 122, 123, 126
Glines, W. M., 233-34
Gnadenhutten, 38, 55, 58, 63, 67-70
Gnadenhutten Massacre, 39, 40, 64-65, 68
Goshen, 69, 70
Grand River (Ontario), 140
Granger, Gideon, 198
Great Lakes, 41, 77, 120-121
Great Meadows, 7
Great Trail, 30, 32, 39
Green Bay, WI; 125
Greene, Griffin, 173
Greene, Charles, 173
Greene, Nathaniel, 173
Greene County, 187
Greenville, 146, 150
Greensburgh, Pa; 223

Greenville Treaty, 68, 107-08, 140, 199, 227
Grouseland, 44-45, 152
Gulf of Mexico, 172, 177
Guyasuta, 10-11, 29
Gwinnet, Button, 31

"**Ha**-Hetuck", 169
Half King, 29, 61, 62
Hamilton, Alexander, 92-93, 110
Hamilton, Henry, 30, 35-6, 102
Hand, General Edward, 59
Hardin, John, 91, 98, 102, 137
Harmar, General Josiah, 16, 84, 98-101, 137-38, 149
Harper's Monthly, 235
Harrison, Benjamin, 36
Harrison, William Henry, 20, 36, 44-46, 111, 151-3, 155-6, 191-2, 206
Harvard College, 197, 223
Havana, 114, 177, 179
Hay, John, 114
Heckewelder, David, 50
Heckewelder, Regina, 50
Heckewelder, John, 49-72, 80, 103
Heckewelder, Sara Ohneberg, 61, 65, 71
Henry, Patrick, 36
Hinckley, 203
Howe, Henry, 209
Hubbard, 203
Historical and Philosophical Society of Ohio, 196, 207

History, Manners and Customs of the Indian Nations Who Once Inhabited Pennsylvania and Neighboring States, 71
Hocking River, 27, 190
Hopkins, Esek, 162
Hopkins, Stephen, 162
Hudson, David, 201-203
Hudson River School, 207
Hull, General William, 156
Huntington, Samuel, 204
Hutchins, Thomas, 65

Illinois, 30, 127, 187
Independent Chronicle, 188
Indiana, 141, 146
Indiana Territory, 20, 151
Iroquois tribe, 35, 40, 119, 129, 131-132, 137, 139, 141
irvine, General William, 39
Irving, Washington, 114

Jackson, Andrew, 111, 114-115, 210-211
Jamaica Fleet, 166
Jefferson, Thomas, 26, 27, 36, 44, 109-10, 112-14, 178, 190-92, 198
Johnson, Richard Mentor, 158
Johnson, Sir William, 127, 130, 132
Johnson Hall, 131-32, 140
Jones, John Paul, 165

Kanawha River, 2, 78, 81, 85
Kaskaskia, IL; 30
Kellog, 200, 201-03

Kenton, Simon, 25, 30-31, 145, 150
Kentucky, 14, 24, 30, 94-97, 141-43, 149, 176, 182, 189, 190, 193
Kentucky Gazette, 182
Kiashuta, *see* Guyasuta
Kenyon College, 208
Kinsman, 203
Kirker, Thomas, 150
Kirtland, 203
Knight, Dr. John, 39-40
Knox County, 226
Knox, Henry, 66, 138
Kokosing River, 226

Lake Champlain, 68, 91
Lake Erie, 121, 124, 157, 170, 201, 227
Lake Ontario, 201
Lancaster, 190
Lancaster, Pa; 20, 21
Land Act, 192
Land Ordinance, 13
Lark, 165
LaSalle, 3
Laurens, Henry, 33, 58
Laws of the Territory of the United States Northwest of the Ohio, 186
Lebanon, CT; 131
Lee, Arthur, 166
Leominster, MA; 218, 236
Lewis, Colonel Andrew, 24
Meriweather, Lewis, 73, 111, 113, 115, 173
Lewisburg, WV; 76-7

Gliding to a Better Place

Lexington, KY; 182
Lexington, MA; 64, 225
Lichtenau, 60
Licking County, 226
Licking River, 216
Limestone (Maysville), KY;189
Lincoln, Abraham, 213, 217
Lincoln, General Benjamin, 167-68
Lindsay, Vachel, 236
Little Eagle, 22
Little Miami River, 141, 187
Little Turtle, 99, 100, 138, 139, 145, 146, 149
Liverpool, England; 74
Liverpool (horse) 78, 81-82
Lochry, Colonel Archibald, 37, 100
Logan, 19, 20, 23, 24-6, 28
Logstown, PA; 22, 102
London, England; 51, 82, 128, 132, 135, 161, 183
Long Knife, 22
Longmeadow, MA; 221
Louisiana Purchase, 109, 112
Louisiana Territory, 110, 115
Louisville, KY; 176
Luther, Martin, 50

MacArthur, Duncan, 150
Mad River, 142
Madison, James, 48
Mamete, 147
Manifest Destiny, 14
Mansfield, 226-27, 229, 232, 236
Mantua, 201

Marietta, 15-16, 41, 65-67, 69, 73, 98, 134, 136, 160, 169, 172-73, 178-79, 184, 192, 214, 216, 222
Marshall, John, 113, 116
Martin, Luther, 113-14
Martinsburg, WV; 188
Maryland, 89, 188
Massachusetts, 187, 196
Massie, Nathaniel, 190
Maumee Rapids, 105, 137, 139, 156
Maumee River, 68, 106, 119, 127, 137, 138, 229
Maxwell, Nancy Robins, 185-187, 189
Maxwell, William, 182-187
Maxwell's Code, 186
McGuffey's Reader, 26
McIntosh, General Lachlan, 31-33, 35
McKee, Alexander, 29, 58
Meigs County, 11
Meigs, Return Jonathan,156, 170
Mercer County, 100
Mexico, 111, 112
Mexico City, 117
Miami River 30, 37, 97, 144, 190, 192
Miami tribe, 136, 149
Mingo (or Seneca) tribe, 9, 22, 23, 29, 31, 34
Mingo Junction, 9
Mississippi River, 3, 42, 94-5, 108, 161, 172, 176
Missouri, 141, 145

Mohawk tribe, 51, 129-30,140
Mohawk River, NY;132-33,201
Mohican River, 226
Moluntha, 144
Monongahela Farmer,172,176
Monongahela River, PA; 172
Montreal, 68, 120, 131
Moravia, 50
Moravian Church, 19, 33-34, 38, 52-57, 60-65, 68, 70-71
Moraviantown, Ontario, 158
Morgan, Daniel, 25, 91
Morris, Thomas, 211
Morristown, 226
Mount Blanchard, 229
Mount Vernon, VA, 9
Mount Vernon, 226-27
Muskingum River, 15, 35, 41, 55, 84, 98, 99, 136, 160, 169, 170, 190, 216, 226

Nantes, France, 165
A Narrative of the Mission of the United Brethren among the Delaware and Mohegan Indians, 72
Natchez, MS; 84, 113, 176
Nathez Trace, 189
National Education Association, 208
National Road, 6
The Navigator, 178
Netawatwes, 23,31, 54-5, 59
New Gnadenhutten, MI; 64-5
New Hampshire, 131
New Harmony, IN; 207, 208
New Jersey, 110, 182

New Madrid earthquake, 155
New Orleans, LA; 42, 59, 94-6, 101, 109-10, 113, 115, 117, 137, 171-72, 176-77
New Philadelphia, 55, 70
New Schoenbrunn, 64
New York, 29, 35, 68, 109, 116, 129, 134, 201
New York City,NY; 54, 59, 110, 128, 183, 189, 198, 207, 212
Newburgh Petition, 12
Newfoundland, 166
Newcomerstown, 54
Niagara Falls, 68, 125, 156, 201
Niagara River, 201
Northampton, MA; 197, 200
Northwest Territory, 13, 15, 41, 43, 78, 101, 136, 169, 170, 199, 200
Norwalk, 224
Notes on the State of Virginia, 28

Oberlin College, 212
Ohio Company, 13, 15, 65, 79, 169, 185, 225
Ohio Company of Virginia, 4
Ohio River, 2-5, 15, 23, 25, 30, 35, 37, 55, 63, 67-8, 75, 78-9, 85-6, 111, 135, 139, 160-61, 178, 182, 189, 207
Ohio University, 13
Ohio Valley, 11, 19, 28, 47, 74
Ojibway tribe, 120
Old Britain, 4
Oneida tribe, 132

Gliding to a Better Place

Ontario, 66
Ordinance of 1787, 12-14, 184, 191
Oriskany, Battle of, 133
Osage tribe, 153
Ottawa tribe, 22, 29, 120
Owen, Robert, 207

Paris, France, 78-79, 128, 165
Parkersburg, WV; 111
Parsons, Samuel, 15, 185
Pauli, Ensign Christopher, 124-125
Pennsylvania, 8, 10, 15, 24, 28, 29, 35, 38, 41, 52, 54, 93, 178, 190, 199, 226
Perry, Oliver Hazard, 157
Philadelphia, PA; 43, 53, 59, 67-68, 89, 110, 128, 138, 178-179, 183, 189, 230
Philadelphia Medical and Physical Journal, 70
Philanthropist, 208
Pickawillany, 4
Pike County, 193
Pipe, Captain, 52, 58-62
Piqua, 4, 143
Pittsburgh, PA; 9, 11, 20, 24, 30, 39, 40, 43, 47, 54, 66, 79, 110, 176, 207
Pittsburgh Gazette, 182
Playfair, William, 78-79
Point Pleasant, 25, 31, 75, 78, 81, 85, 87, 100, 142-143
Pontiac, 22, 29, 45, 97, 106, 119-127
Pontiac's Conspiracy, 10, 19, 22, 52, 53, 102, 122-27, 148
Port Washington, 61
Portage County, 196, 200, 201, 204
Posey, Thomas, 47
Post, Christian Frederick, 51-4
Powhaten Point, 10
Princeton College, 60, 110
Procter, Henry, 156-157, 158
The Prophet (Lalawethika), 45, 147-49, 151, 155
Prophetstown, IN; 45, 151, 155
Providence, 165
Providence, RI; 161, 163, 167, 169
Publick Occurrences, 182
Put-in-Bay, 158
Putnam, General Rufus, 12, 15, 65-69, 79-80, 102

"Quakers", 162
Quakers, 53
Quebec, 8, 52, 137
Queen Anne of England, 74, 82

Rabelais, 209
Randolph, John, 113
Ravenna, 196, 203-05
Ravenna Township, 200
Reno, 10
Reports on Cases Decided in the Courts of Common Pleas of the Fifth Circuit of Ohio, 206
Richmond, VA; 36, 113
Rhode Island, 163, 164, 168

Rogers, Major Robert, 120-21, 126
Rose, 165
Ruddell, Stephen, 143, 150
Rumford, Count, 219

Sabine River, LA; 112
Sacagawea, 73
St. Clair, 160, 173, 175-78
St. Clair, Arthur, 15, 16, 24, 41, 65, 84, 91, 99-101, 136, 138, 149, 173-74, 184-85, 191-92, 204
St. Clair's Defeat, 42, 100, 105, 136, 138
St. Clair, Arthur, Jr., 191
St. Clair, Louisa, 136
St. Clairsville, 193, 207
St. Francis of Assissi, 218
St. Lawrence River, 68, 125
St. Louis, MO; 110
St. Petersburg, Russia, 178
Salem Mission, 61
Sandusky, 134
Sandusky Plains, 39
Saratoga Campaign, 61, 91, 93, 98, 133
Sargent, Winthrop, 15, 184-85
Saxony, 50
Schoenbrunn, 55, 69
Scioto Company, 78-79
Scioto Gazette and Chillicothe Advertiser, 188-93
Scioto River, 142, 190
Seneca tribe, 129
Shawnee tribe 24, 25, 29, 31, 45, 59, 75, 132, 136, 141-45, 149, 150
Shepherdstown, WV; 188
Six Nation Confederacy, 31, 129, 132, 134
John Smith, 110, 114
Smithsonian Institution, 196, 213
Souix tribe, 153
Spain, 95 101, 108-10, 112
Spanish Conspiracy, 116
Springfield, 142-43, 150
Sproat, Ebenezer, 168-69
Society for the Propogation of the Gospel Among the Heathen, 65
Stanton, Edwin, 196, 213
Staunton, VA 75
Steubenville, 23, 205- 08, 213, 222
Steubenville Western Star, 206
Stow, 203
Stirling, General William, 92-93
Struthers, 203
Stuart, Gilbert, 132, 196, 198
Suffield, CT; 198
Sullivan, General John, 188
Susquehanna River, PA; 54
Swedenborg, Emmanuel, 218, 223
Swedenborgian Society, 230
Swift, Jonathan, 209
Symmes, John, 185

Taney, Roger B. , 116
Tallmadge, 203
Tappan, Arthur, 212

Tappan, Benjamin, 195-213
Tappan, Benjamin, Sr. 197, 200
Tappan, Benjamin, III, 208
Tappan, Betsy Frazer, 208
Tappan, Eli Todd, 208
Tappan, Lewis, 212, 213
Tappan, Nancy Wright, 204, 205, 208
Tappan, Sara Homes, 197
Tappan's Reports, 206
Tarhe, 149
Taylor, Major Richard, 34
Taylor, Zachary, 47
Tecumseh, 20, 45-46, 71, 119, 140-159
Tennessee, 14, 145
Terre aux Boeufs, LA; 115
Texas, 109, 212
Thames River, England;161, 168
Thames River, Ontario;158
Thayendanegea, 129
Thompson, Benjamin, 219
Tiffin, Edward, 190, 193
Tippecanoe Campaign, 20, 46, 100, 155
"Tippecanoe and Tyler, Too", 158
Tippecanoe River, 45
Toledo, 137
Treaty of Fort Harmar, 136
Treaty of Fort Wayne, 151
Treaty of Paris, 8, 102, 105, 126, 134, 139, 168
Trotter, Richard, 75
Trotter, William, 75, 85-86

Trumbull County, 204
Tu-Endi-Wei Park, 87
Turner, Frederick Jackson, 89
Turner, George, 185
Tuscarawas River, 18, 32, 35, 38, 51, 52, 54-55, 60, 63, 70, 127, 135
Tuscarawas County, 70

Uncle Tom's Cabin, 225
Unitarianism, 197
Upper Sandusky, 61, 62, 64
Urbana, 150, 229

Valley Forge, PA; 215
Van Every, Dale, 4, 5
Van Wert, 229
Varnum,James, 15, 185
Vermont, 14
Vincennes, IL; 30, 35, 44-45, 47, 66, 67, 96, 103, 151-52, 154, 192
Vinez, Elizabeth de, 21
Virginia, 3, 8, 24, 28- 30,36, 44, 75, 78, 94-95, 113, 193
Virginia, University of, 14
Voltaire, 7
Von Steuben, Baron, 36

Wadsworth, 203
Wadsworth, Elijah, 205
Walhonding River, 226
Wallace, George, 47
Wapakoneta, 149
Warren, 202
Warren, PA; 222
Washington, D.C., 115-116,

152, 189
Washington, George, 1-17, 29, 36, 47, 92-93, 98, 107, 128, 138, 166, 187, 191, 198
Wayne, Anthony, 16, 68, 88-89, 101-103, 105, 107-9, 139-40, 146, 150
Wells, Bezaleel, 210
Wells, William, 66, 149-50, 156
Wesley, John, 50
West Indies, 51, 160
West Liberty, 144
West Virginia, 84
Western Reserve, 199-200, 203, 204
Western Reserve College, 202
The Western Spy and Hamilton Gazette, 175, 187
Wetzel, Lewis, 83-84
Wetzel, Martin, 84
Wheeling, WV; 38, 62, 83, 185-6
Wheelock, Eleazor, 131
Whipple, Abraham, 160-186
Whipple, Catherine, 168
Whipple, John, 169, 170, 177
Whipple, Polly, 169
Whipple, Sarah Hopkins, 162,179
Whiskey Rebellion, 20, 42-44
White Eyes, 29, 32, 59-60
White River, IN; 146
Whitney, John, 233
Whitney, Sally Chapman, 216, 233
Wilkinson, Ann Biddle, 93 110, 112

Wilkinson, James,66, 88-117
Williams, Roger, 161
Williamsburg, 5, 145
Williamson, David, 39, 63, 64, 67
Willis, Lucy, 193
Willis, Nathaniel, 188-194
Willis, Nathaniel, Jr., 193
Willis, Nathaniel Parker, 193
Wilson, James, 206, 209
Wilson, Woodrow, 206
Winchester, VA; 188
Winship, Winn, 188, 190
Woodbridge, Dudley, 113, 173, 175
Woodbridge, Dudley, Jr., 173
Worthington, Thomas, 111, 150, 151, 188, 190-192, 206
Wounded Knee, 100, 148
Wyandot tribe, 29, 34, 58, 61, 62, 63, 149
Wyoming Valley, PA; 133, 199

Xenia, 141, 146

Yorktown, 62
Youghiogheny River, 7
Young, John, 223
Youngstown, 202

Zane family, 83, 186
Zane, Betty, 186
Zane, Ebenezer, 186, 189-90
Zanesville, 136, 190, 207
Zane's Trace, 189-190, 193
Zeisberger, David, 33, 34, 54, 55, 57, 64, 66, 68, 69, 71